Practical Atlas for Bacterial Identification

SECOND EDITION

About the Cover

Growing rusticles (3, 19-26 RST) became important after researches of the wreck site of the *RMS Titanic* off the continental shelf in the North Atlantic. The exposed steel surfaces of ship appeared to be coated with rusticles. In 1996 and 1998, expeditions recovered rusticles from the ship and by late 1999 methods were found to culture these consormial growths in a laboratory. The picture shows the manner in which rusticle growth may be enhanced using electrically impressed steel plates. In 2000, the technology advanced to the point where an aquarium of cultured rusticles went on tour with the Titanic Science exhibition organized by the Maryland Science Center. Over the next five years, more than 2.5 million exhibition visitors witnessed the growing rusticle consorms that formed in seven distinct growth patterns.

Practical Atlas for Bacterial Identification

SECOND EDITION

D. Roy Cullimore

CRC Press
Taylor & Francis Group
Boca Raton London New York

CRC Press is an imprint of the
Taylor & Francis Group, an **informa** business

CRC Press
Taylor & Francis Group
6000 Broken Sound Parkway NW, Suite 300
Boca Raton, FL 33487-2742

First issued in paperback 2019

ISBN-13: 978-1-4200-8797-0 (hbk)
ISBN-13: 978-0-367-38443-2 (pbk)

Library of Congress Cataloging-in-Publication Data

Cullimore, D. Roy.
    Practical atlas for bacterial identification, 2nd. Ed. / D. Roy Cullimore.
      p. cm.
    Includes bibliographical references and index.
    ISBN 978-1-4200-8797-0 (hardcover : alk. paper)
    1. Bacteria--Identification--Atlases. I. Title. II. Title: Bacterial identification.

QR54.C85 2010
614.5'7012--dc22
                                          2010004390

Visit the Taylor & Francis Web site at
http://www.taylorandfrancis.com

and the CRC Press Web site at
http://www.crcpress.com

Dedication

In Memory of Bob Woolsey
Visionary and Thinker
Scientist and Engineer
Listener and Believer
Still Alive in the Memories of the Many

Contents

List of Tables

List of Figures

List of Plates

Preface

Welcome to 1830—a time when Europe was recovering from the Napoleonic conflict and Napoleon left a legacy that included canning meats to control spoilage by the application of heat. Nicholas Appert was credited with this discovery of microbiologically influenced spoilage of meats. The ability to control spoilage helped the French-led troops to reach the outskirts of Moscow before the Russian winter played its hand. Canning was the first major advance in the application of heat to control microorganisms.

Another battle in progress then continues today. In 1830, four science factions were fighting for dominance over the emerging field of microbiology. The botanists and zoologists were two major contestants who already controlled the plant and animal kingdoms; microbes were considered members of one of the two kingdoms. Major progress in chemistry led to the understanding of the natures of chemical elements and the complexities of life processes. Chemists, in conjunction with the then-primitive medical and pharmaceutical sciences believed that all diseases were caused by chemicals called miasmas. The fourth major contestants were the western religious movements that considered microbes the work of the devil based on the belief that the human eye was capable of seeing every living thing. Microbes were too small to be seen except with a compound microscope and thus the "invisible" microbes were the devil's creations. Any scientist who investigated microbes sinned by investigating the work of the devil.

From 1830 to 1870, microbiology did not develop significantly because of the negative forces exerted by religion (unseen entities were the works of the devil) and chemists (all significant events are chemical in form). Positive reinforcement for the development of microbiology came from the botanical and zoological scientists, but there were caveats. First was the dogmatic application of the Linnaean reductionist concepts: all life forms were independent species that were not dependent on other species. These concepts remain in effect today and microbial species are judged only when they can be cultured independently. No consideration was given to the possibility that communities of microorganisms participated in environmental activities involving soils, waters, sediments, crustal elements, and clouds.

The years from 1850 to 1880 saw development of intensive monocropping followed by the major fungal infestations, primarily in the form of cereal rusts that decimated crop yields. Plant breeders rapidly developed rust-resistant strains of cereal plants and the fungi were recognized as major plant pathogens and thus embraced as plants by the botanists.

Louis Pasteur undertook a sequence of scientific investigations that primarily demonstrated that microorganisms resulted from reproduction by living cells and not from activities of the devil. Pasteur's proof finally put to rest the debate that spontaneous generation of microorganisms was the work of the devil. Religious objections faded, leading to private and government sponsorship of research that gave birth to the golden age of bacteriology from 1880 to 1890.

Zoologists and botanists eventually allowed bacteriology to be recognized as a science, but identification of bacteria continued to follow the reductionist Linnaean form of classification. Breakthroughs occurred almost weekly as typhoid, dysentery, and cholera were attributed to the work of specific bacterial strains. A landmark in this sudden spurt of science was a suggestion by Frau Hess, the wife of a bacteriologist, to use an agar-based jelly to grow bacteria. She made the suggestion at a time when several major bacterial pathogens were under investigation. Agar provided the most effective means for culturing bacteria but excluded bacteria that would not grow on agar-based media. Reductionists seized on the superiority of agar and ignored the many bacteria that cannot grow on agar. Today the science of bacteriology is dominated by reductionist thinking and the premise (in common with botany and zoology) that any microorganism must be cultured as pure strain, free from contamination by other strains.

During the 20th century, bacteriology underwent a number of changes arising from advances in chemistry and medicine. Two breakthroughs in chemistry were the recognition of the immune system as an organic chemical neutralization process and the development of the sulfonamide drugs to control infestations. Observations of antibiotics generated by microorganisms (particularly fungi) were reported from the start of the century, then ignored. In the late 1920s, Sir Alexander Fleming developed penicillin, but it was ignored in favor of sulfonamides until World War II created an urgent demand for controlling wound infections. This demand brought penicillin out of the closet and it became a major life saver for wounded troops. After the war, the focus was on developing more antibiotics from natural sources and eventually led to molecular manipulation of the classic antibiotics and today most antibiotics have been refined through biochemistry and genetics.

In the 1950s a chemical tsunami swept through all the biological sciences with the discovery of the chemical structures of nucleic acids, most notably deoxyribonucleic acid (DNA). This discovery allowed precise classification and categorization of genetic material. In bacteriology, the advance led to frenetic reductionist activity as the genetic and biochemical aspects of selected strains were developed. Lost in this research stampede were the bacteria perceived as insignificant: they did not grow on standard cultures or they functioned collectively with other strains only within community structures. They were considered inconspicuous, irrelevant, and unworthy of attention. Who would care about a crust of rust, a mound of slime, or a glistening sheen? The 21st century is another reductionist era in bacteriology when all bacterial functions occur within a cell wall bag of chemicals dominated by nucleic acids, enzymes, and energy drivers.

Most environmental settings present two issues that must be addressed, The first is that the functioning of each bacterial cell is influenced by the heterogeneous community within which it functions. Such communities commonly involve 8 to 64 individual strains of bacteria, all performing different tasks in an integrated and intelligent manner. The second issue is the dispersal of cells in water that is bound by an extracellular polymeric matrix that creates volumes that exceed the volumes of incumbent cells by two to three orders of magnitude.

Maturation profoundly influences the heterogeneous nature of this bacterial growth. Maturation generally involves reduction of the numbers of active bacterial strains present, reduction of the bound water surrounding the incumbent community, and increases of bioaccumulated materials "stockpiled" within the bound water. The result is that a bacterial population active within a growth declines and incumbent cells may assume states of suspended animation (such as those generated by endospores or ultramicrobacterial cells). When this state is reached, a growth may contain no active bacteria but continue to support large populations of bacteria that are in a resting (suspended animation) state.

For the applied microbial ecologist seeking to identify bacteriologically influenced growth, one requirement is the precise identification of the observed growth based on holistic approaches, not on cultured isolates that can create considerable bias. This second edition of the *Practical Atlas for Bacterial Identification* has been written to provide a means to categorize bacterial growths into larger heterogeneous communities based on form, function, and location of growing biomass. No attempt is made to define individual species; doing so would require manipulation of the biomass of interest to remove samples for biochemical or sequential cultural analysis and thus affect the value and utility of the data generated. This atlas has been constructed as a guide to the identification of the principal consorms (integrated bacterial communities) in specific environments, with emphasis on the forms, function and locations of the definable consorms.

The first (2000) edition of this atlas was based on the premise that all the major bacterial genera could be located on two-dimensional maps that would show the genera as "cities" and groups of like genera as "countries." No formulation for the grid that would allow positioning with a greater level of precision was included in the first edition. The second edition includes a bacteriological positioning system in which the oxidation–reduction potential (in millivolts) serves as the longitudinal (x) axis and the viscosity of the water within the consormial biomass measured in log centipoise represents the latitudinal (y) axis—an approach that allows positioning of each consormial biomass within a grid system. This edition focuses on differentiating the major consormial biomass groups recognized by observation or speculated upon based on the recognition of their diversity and durability within the various environments of the biosphere.

This edition contains 12 chapters and an appendix to guide readers through this approach to the identification of bacterial consorms. It follows the less popular practice of "lumping" groups of bacteria together rather than the more expedient and popular "splitting" of bacteria into the smallest grouping that allows activity and growth to occur. This second edition is designed to be used by practitioners who design and manage engineered systems and find that bacteria are the causes of some of the economic and sustainability challenges, whether chemical, metallurgic, or physical. For example, corrosion is often caused by bacterial consorms that either generate acids or create electrolytic conditions that compromise materials of concern. In groundwater, consormial biomasses tend to concentrate around the oxidative–reductive interface and cause clogging if sufficient bioaccumulations are present or plugging if higher water content obstructs water flows.

Chapters 1 and 2 deal with the locations and functions of bacterial communities that can form consorms and the common events they initiate within natural or engineered ecosystems. Chapter 3 defines six major alpha groupings of bacteria consorms and generates the concepts employed in the gridded atlas positioning system used to differentiate the major bacteria consorms. Chapter 4 expands on the differentiation described in Chapter 3 and details the positioning of major defined bacterial consorms. Chapter 5 examines the environmental dynamics that influence the manner in which bacterial consorms will develop. Chapter 6 addresses the challenges in defining the forms, functions, and habitats of the consorms categorized and listed in Chapter 7.

Chapter 8 covers biochemical methods used to identify consorms. Cultural identification methods based on bacteriological activity reactions are addressed in Chapter 9. Explanation of the gridded atlas location system is covered in Chapter 10. Chapter 11 discusses linkages of the bacterial consorms and bacteriologically influenced events such as plugging, corrosion, biogeneration of natural gases and oils, and natural systems. Chapter 12 illustrates consormial events by specific examples. The appendix is an update of Chapter 10 from the first edition and includes the original format for alpha two bacteria presented as genera with no attempts to map them into consormial structures.

In 1830 when bacteriology was gradually dominated by reductionist thinking, Linnaean concepts relevant to classifications of plants and animals were employed for bacterial identification. These concepts have only limited value for identifying bacteria that are active in consorms. This atlas goes back to that time because that was when the concept of intelligent, coordinated bacterial consorm activity was lost in the 100-year stampede to find the causes of bacteriologically influenced diseases, but only if they grew easily on agar plates! How many diseases are consormial in origin? How many more have not been discovered because they cannot be cultured by agar methods? Bacterial consorms are visible. Their locations tell a story, and their forms and functionalities allow identification without the need to isolate a single species of bacteria.

This atlas contains no references, footnotes, or endnotes simply because they would turn this book into another boring reference source. The suggestions for further reading may be useful but my hope is that all readers begin to see that bacterial consorms are as evolved as humans to the degree that they act with intelligence to acquire and maintain their homes.

SOME FURTHER THOUGHTS

If I could but sit just still in one spot and let the world surround me with truth, then and only then would I smile. Greyness is the color that I wear. Everywhere will I look for questions for which there are no answers. The only thing about here is now. Beyond this now there lies pent up the forces of the future waiting to be released. Perhaps I am tired of treading the pathways of thoughts running through the thickets of dogma. Perhaps I will never see the harvest from the seeds of science that I did sow. But I say to them, may they grow and swirl out of the soils of uncertainty into pleasant moments of reality then that, for me, will be enough. This story is my story of why things have come to be and why science will always remain that haven for mystery. That is the why I gaze at the sky, ponder into a wall, never kick the soil, and enjoy the pleasures of breathing.

D. Roy Cullimore

Acknowledgments

This second edition of the Atlas is the culmination of the efforts of the staff of Droycon Bioconcepts Inc., Regina, Canada, in examining as we worked many aspects of many different environments exhibiting significant bacterial activities. Their observations were based on common sense, field work, experience, and laboratory investigations. They examined ochres; rusticles from the North Atlantic, Gulf of Mexico, and Mediterranean; rain storms; corroded steels; failing concretes; gas hydrates; groundwaters; biofouling at industrial plants; and spoiling foodstuffs to demonstrate the roles of bacterial consorms.

For the application of common sense and refined analysis I acknowledge the late George Alford who developed the blended chemical heat treatment method for water wells. George's knowledge directed the best approaches to the control, disruption, and effective dispersement of offending biomasses, knowing that these iron-rich growths would reappear unless preventive maintenance was strictly applied.

Also acknowledged is the late Bob Woolsey of the University of Mississippi's Mississippi Mineral Resources Institute and Center for Marine Resources and Environmental Technology. Bob undertook many deep-ocean projects, primarily in the Gulf of Mexico, particularly at MC 118 gas hydrate, and was very open minded about the potential for significant bacteriological events in extreme environments.

The plates in Chapter 12 were produced by the staff of Droycon Bioconcepts Inc. as they faced bacteriological challenges in the various environmental niches subjected to investigations. Staff support, enthusiasm, and common sense helped greatly in preparing this second edition.

Finally, I would like to acknowledge those who recognize the need for a comprehensive guide to identify bacterial consorms and seek simple answers to resolve the relevant issues. Hopefully, this second edition will satisfy the need to better understand the complex, dynamic, and intelligent natures of all bacterial consorms.

D. Roy Cullimore

Author

D. Roy Cullimore[*] was trained in agricultural microbiology at the University of Nottingham. Always innovative and creative, Roy published an article in *Nature* on the genetics of fat synthesis in yeast based on his honors thesis, then chose to move into microbial ecology. His doctoral work included the development of a biologically driven soil fertility test that was later patented. Roy's passion for microbial ecology led him from soils to groundwaters and then on to the deep oceans. While investigating water wells, he helped uncover the complex and challenging communities of bacteria that cause severe biofouling risks. These threats applied to both human health and to the health of the water wells.

In the 1980s and 1990s Roy developed diagnostic systems for the determination of the bacterial consorms that cause plugging and corrosion problems in water wells. His diagnostic research led to the invention of the Biological Activity Reaction Test (BART™) that has been patented and is now commercially manufactured. At the same time, he developed rehabilitative/regenerative treatments for wells with coinventor George Alford. Today more than 6,000 major water wells of all types benefit from sustainable production as a result of that work.

One principal target was the iron-related bacterium that causes wells to fail—the same one active on the *RMS Titanic*. Since 1996, both short- and long-term experiments were conducted on many shipwrecks to determine the durability of ship steels under deep-ocean conditions.

For the past 10 years, Roy has been editing a series of books on sustainable water wells and has developed the bacterial atlas concept; the first edition was published in 2000. The atlas concept is expanded in this second edition through the inclusion of more bacteria, primarily as consorms, and the development of a bacterial positioning system that allows placement of all bacterial consorms within the atlas grid. Roy operates Droycon Bioconcepts Inc. as the BART manufacturing operation along with a research and development wing that focuses on applied microbial ecology and consulting.

---

[*] This photograph was taken by Jeremy. B. Weirich, National Oceanographic and Atmospheric Administration (NOAA) while Dr. Cullimore was diving to the *RMS Titanic* in 2003 from the *R/V Keldysh*, a MIR submarine operated by the Russian Academy of Science.

1 Bacterial Communities by Location and Function

1.1 INTRODUCTION TO LAYERING OF BACTERIAL COMMUNITIES

Life in its most basic form occurs in four bands around the Earth. Of the four bands, we are most familiar with the second band that forms the surface biosphere and includes virtually all plants and animals as well as many of the bacterial communities that humans perceive to be important, However, the bacterial communities dominate the other bands of life—often virtually unseen and mostly unrecognized. The four bands of life are easily defined from the outermost to the innermost by their positions relative to the solid or liquid surfaces of the planet, whether in mountains, soils, or deep under the oceans.

For the outermost layer of life, this primarily bacterial biomass occurs in a dispersed state above the solid or water-driven surfaces of the Earth. It is perhaps an ironic twist of fate that we all recognize a certain biomass when it floats in the atmosphere; we call it a cloud. Very few recognize clouds for what they are: floating, dispersed, very active bacterial communities. The common explanation is that water has nucleated around dust particles in the atmosphere. The reality is that bacteria, through their strong ability to bind water, manipulate electrical charges, and control density, can float in the atmosphere as bioparticles. The bionucleated clouds of bound water particles are very reactive within that environment.

In some ways, clouds can be viewed as Nature's floating Band-Aids because the bacteria within the clouds intercede with the movements of volatile chemicals and particles moving upward into the atmosphere. Clouds therefore serve as influences that control the pathways through which chemicals and particles move upward into the atmosphere from soil and water surfaces. The net effect of this intercession is that clouds reflect the levels of burdens generated on clouds by these vertical migrations. An examination of cloud formations over regions that heavily pollute the atmosphere shows that clouds become denser and discoloured due to great accumulations of these pollutants.

Below the first layer of dispersed bacterial floating biomass, a second layer of bacterial activities is found in the surface biosphere that is rich in diverse forms of life including all plants and animals that generate or utilize organic matter that then serves as a focus for microbial activities through two processes: pathogenesis (generation of disease) and saprogenesis (production of putrefaction and/or decomposition). These two processes function at different times. Pathogenesis occurs in a living host; saprogenesis usually occurs after the death of a host but also during decomposition of waste materials excreted from animals.

1

Within the surface biosphere, bacteria are viewed as pathogens causing specific diseases within economically important species or saprophytes causing decomposition of organics. While the role of bacterial pathogens has been recognized particularly as a result of direct interaction with the human species (e.g. typhoid, cholera, anthrax, and dysentery), this recognition is limited to single-strain pathogenic communities that can grow and be isolated through agar culture. It is probable that many bacterial pathogens affecting humans are yet to be discovered by the dogmatic dependence on agar-based isolative techniques.

The discovery two decades ago that gastric ulcers are caused by bacteria (Helicobacterium species) occurred via happenstance: cultures were left to incubate longer than was customary. Perhaps many conditions considered carcinogenic (malignant and uncontrolled growth of tissues) may in fact be the results of influence by bacterial infective agents.

While the surface biomass may appear to be dominated by plant and animal life upon soils and in waters, little attention has been paid to the microbial biomass that significantly dominate the total biomass within the surface biosphere. Movement and size above the ground lead to a natural impression that the plants and animals do indeed dominate, at least in the lines of sight.

Beneath the second layer of biomass dominated by the presence of oxygen lies the third layer of bacterial biomass that is perhaps blessed by being out of sight and out of mind. This biomass stretches down from the fringes where oxygen disappears (oxidation–reduction interface) into the increasingly reductive water-saturated regions of the porous and fractured crust. The groundwater saturating these formations generally shifts downward in salinity from low values near the surface to saturation at depths. Near-surface groundwaters are often of potable quality and frequently extracted as major sources of water.

In addition to the increase in salinity, a second front forms as products of the organic blooms from the surface biosphere move down into groundwater. The bacterial biomass creates a number of natural barriers, with the most intense region of activity at the oxidation–reduction interface. It is here that the biomass often becomes significantly active to cause: diversion or plugging of groundwater movement; filtration of significant organics and chemicals from the water, leading to changes in groundwater quality downstream; and electrolytic and acidulolytic effects, leading to the initiation of corrosive processes.

Beneath the oxidation–reduction interface then conditions become increasingly reductive (oxidation–reduction potential [ORP] declining from –50 to –600 millivolts [mv]). The role of these electrolytic forces at these depths is not yet understood, but it does appear that a smooth transition from bacteria that dominate primarily at the oxidation–reduction interface and the bacteria below occurs. At these greater depths, the bacteria become aggressively reductive, fundamentally stripping various nutrients (phosphorus, nitrogen, and sulfur) from the descending organics and leaving various hydrocarbon (C_xH_y) compounds, primarily in the form of methane and petroleum. Evidence also indicates that bacteria may have been involved in the reduction of organics to more elemental forms of carbon through the formation of various grades of coal (e.g., peat under surface acidic conditions, lignite, and coals). This process of reducing the descending organic fronts moving down into the crust involves two primary steps: (1) stripping of nutritionally usable elements; and (2) generation of hydrocarbons within gas reserves and pools of crude oil locked in deeper geologic regions.

There is an old adage that "what goes down then comes up" and this applies to the completion of the organic carbon cycle within planet Earth. Reduced forms of carbon as gases and hydrocarbons may now move upward through the crust as volatiles, gases, or seepages of the lighter density fractions of oils. Another interfacial layer of biomass forms and follows one of two alternative functions: (1) volatile and gaseous hydrocarbons along with seepage are broken down at the oxidation–reduction interface to generate biomass; or (2) gaseous hydrocarbons (specifically methane) are accumulated at the oxidation–reduction front along with ice to form energy-rich gas hydrates (clathrates).

As the reduced forms of carbon move back into an oxidative zone created by molecular oxygen or by the electrolysis of water, there is the potential for oxidation of these carbon forms back into organic carbon. Many of the bacteria naturally active in the surface biosphere and in subsurface layers of the bacterial biomass may also become active, utilizing the feedstock of reduced forms of carbon emanating from the deeper aggressively reductive zones of the biomass.

Of these activities, the generation of the gas hydrates, primarily at the mud line along the continental shelf and in the tundra, reveals some unique conditions. Methane primarily becomes accumulated along with ice (solid water) commonly at a ratio of 8:1 (methane: ice). These hydrates form in manner resembling icebergs but with two major differences: (1) although temperatures are above freezing, ice is present; and (2) the density of the gas hydrate is such that biomass does not float up in the manner that icebergs do. As reserves of methane (according to the U.S. Geological Service), the gas hydrates represent twice the combined energy reserves of oil, gas, and coal. Methodologies for the successful extraction of the methane gas from the gas hydrates remain to be developed.

Deeper down in the crust, where liquid water permeates downward under higher pressures and meets the very high temperatures emanating from the magma, flash boiling converts the liquid water to steam. While the liquid water could still support life under even these high pressures, the conversion of that water to steam is considered to create a negative environment for sustenance of bacterial activity. It may be conjectured that potential for bacterial activities exists at the interface between the liquid water and the generated steam. This interface would be very unstable and involve considerable fluctuations in temperatures, pressures, and water forms. It is theoretically possible that bacterial life could exist outside the regions of extreme instability where the water, although very hot and pressurized, remains in liquid form. Under such environmental conditions, liquid water would remain relatively stable outside the zones of radical instability. It is conceivable that bacteria, because of their simple cellular construction, could easily tolerate the high pressures and perhaps adapt to the high temperatures. Discovery of these bacteria would add significantly to our understanding of the adaptability and durability of the bacterial kingdom.

1.2 FACTORS SIGNIFICANTLY INFLUENCING BACTERIAL ACTIVITIES AND NUTRIENT CYCLES

Six potentially major differentiating environmental conditions affect the types of bacterial consorms that can become active and grow. The resulting biomass "layers" are very much reflections of the dynamics involved and are often dominated by the bacteria. Perhaps dominant are the influences of bacteria on the movements of key elements in the sustenance of all known forms of life. These dominant elements include carbon, oxygen, hydrogen, nitrogen, phosphorus, and sulfur. They all cycle through the crust and atmosphere of the Earth and their passages are influenced or controlled by bacterial activities.

Each of these elements has a distinctive cyclic pattern but they are interwoven, making understanding of the bacteriological influences that much more challenging. Each element is affected by oxidative (atmosphere and surface biosphere) or reductive (crust) conditions. Often, the interface between reductive and oxidative conditions is where most bacterial activities occur. This is where the surface biosphere is present on the land. In the oceans, this oxidative–reductive interface extends down beyond the photic zone commonly to the mud and sediment zones on the floors of the oceans.

Complex interactions among the carbon, oxygen, hydrogen, nitrogen, phosphorus, and sulfur elements make conceptualization more challenging. Perhaps the best example of this interaction is water (H_2O), composed of two atoms of hydrogen and one atom of oxygen and forming the universal solvent that creates a binding matrix for all life. Not only is water essential to life on Earth but of its two elements, oxygen dominates the oxidative zone and hydrogen dominates the reductive zone. The oxygen and hydrogen dominances meet at the oxidation–reduction interface, generating a constant and measurable electrical output measured in millivolts that are positive in the oxidative environment and negative in the reductive. It could be conjectured that the oxidation–reduction interface commonly located in water provides a natural focal point for electrolysis. This electrolytic activity could drive oxygen upward eventually into the atmosphere and hydrogen downward into the saturated crust.

The hydrogen and oxygen released from the water interact with living systems and the environment. Oxygen, when reacting with carbon, generates final products as carbon dioxide (gaseous CO_2) and carbonates (crystalline CO_3^-). A major intermediate is biomass ($C_AH_BO_CN_DP_ES_F$ and the process continues until virtually the whole periodic table is incorporated). Hydrogen, reductively in a reaction with carbon, generates hydrocarbonaceous gases (C_XH_Y)as final products, usually with chain lengths of shorter than five and dominated by single carbon atom methane (CH_4) and petroleum hydrocarbons (C_XH_Y) that generally have more than five carbon atoms within the chain (polymer) that can extend to fifty or more atoms, often with complex side chemistries. The origin of these hydrocarbons is the reductive degradation of biomass. A bacteriologically driven fermentative process reduces the biomass ($C_AH_BO_C$) to fatty acids ($C_XH_Y.COOH$) and from there, under reductive conditions, to hydrocarbons (C_XH_Y). This strips all the other elements incorporated into the biomass (summarized as $C_AH_BO_CN_DP_ES_F$) from the descending organic molecules until only carbon and hydrogen are left.

This cycle of carbon, hydrogen, and oxygen appears to involve two very distinct processes, both of which are dependent on the presence of water. Oxidative processes are driven by the presence of oxygen and this generates the diversity of life in the surface biosphere. These processes are propelled by the generation of organics that are, in part, utilized for respiration, resulting in the generation of carbon dioxide and release of energy. Reductive processes are driven by the abundance of organic carbon and hydrogen and increasing shortages of the other elements of life (e.g., $O_C N_D P_E S_F$) so that the biomass, dominated by bacteria, scavenges the descending pool of organics for these elements. This scavenging can only be postulated but it is conceivable that of these four elements ($O_C N_D P_E S_F$), nitrogen, and phosphorus would be scavenged first, followed by sulfur and oxygen.

Nitrogen is well known to follow a cycle of fixation (from nitrogen to ammonium), nitrification (oxidation of ammonium to nitrate), and denitrification (from nitrate to gaseous nitrogen, N_2). Much of the nitrogen is present in the atmosphere and is subjected to bacteriologically dominated cycling (nitrogen fixation brings the element down to the surface biosphere and denitrification returning it to the atmosphere). Nitrogen fixation can also occur under reductive conditions when anaerobic bacteria fix nitrogen. Essentially the nitrogen cycle is very dynamic and appears to occur primarily in the atmosphere and the surface biosphere but nitrogen can be expected to become deficient at greater depths.

Phosphorus follows a more complex cycle, primarily in the surface and subsurface biospheres where it appears in four principal forms: (1) soluble inorganic primarily as phosphate or SIP; (2) soluble organic or SOP; (3) particulate organic or POP; and (4) particulate inorganic or PIP. Of these four forms, PIP is the most recalcitrant as crystallized forms of polyphosphates. SIP, SOP, and POP are remarkably mobile within the biomass, cycling between forms in a matter of hours. Most significant of these forms to living biomass is SOP, more particularly the adenosine diphosphate (ADP) and adenosine triphosphate (ATP). The phosphate bonds created in these two molecules are high-energy types and form the major repositories for energy storage within all living active cells. In that sense, phosphorus is essential to all life because ADP and ATP provide a method for storage of high-energy bonds. In the diffusion of organics from the oxidative to reductive zones, it is therefore not surprising that phosphorus is also scavenged quickly.

Sulfur occurs in the surface environment primarily as sulfates in predominantly oxidative zones. Since it is found near the oxidation–reduction interface, the dominant form may shift to elemental sulfur. This is consistent with the green and purple sulfur bacteria that can utilize hydrogen sulfide by generating some forms of elemental sulfur nanoparticles inside or outside cells. Under reductive conditions, bacteria generate hydrogen sulfide from sulfates or sulfur–amino acid-rich proteins. In both cases, the normal recalcitrant products are black sulphides, commonly formed via reactions involving iron. Sulfur therefore cycles between dissolved sulfates (oxidative), through to insoluble sulfides (reductive) with elemental sulfur appearing at the oxidative–reductive interface.

Oxygen is present in the atmosphere and appears to have been derived fundamentally from the photosynthetic functions in plants that reduce carbon dioxide to organic materials; the oxygen surplus from the synthetic function is released as molecular oxygen (O_2). While photosynthesis is commonly accepted as the primary source of oxygen in the atmosphere, oxygen can also be derived electrolytically from water (H_2O) which is abundant primarily as surface and groundwaters and secondarily as biologically and chemically bound water. When electrolysis generates from water as molecular O_2, it requires two molecules of water to generate one molecule of oxygen and two molecules of hydrogen (H_2).

Gravimetrically the electrolysis of water (atomic weight 18) would generate 11% hydrogen and 89% oxygen from each water molecule. Carbon dioxide (atomic weight 28) would theoretically generate 43% carbon to 57% oxygen under ideal conditions. However, much of the generated oxygen would be integrated into the synthesis of organic materials by plants and this would substantially reduce the output of free oxygen by photosynthesis. These ratios indicate that electrolysis generates 89% of the molecular weight of the water as oxygen and is potentially more efficient than photosynthesis that maximally generates 57% of the molecular weight as oxygen.

Two major constraints of the sources of molecular oxygen in the atmosphere are the energy sources triggering the release of oxygen from plants and the molecular efficiency of that release. For oxygen generated from photosynthesis during light periods, the efficiency is compromised by the primary need to utilize the oxygen in organic synthesis and also the need to use the oxygen for respiratory functions during dark periods. Oxygen generated by electrolysis presents no constraints (light and dark periods) on the utilization of oxygen for organic synthesis. Certain factors support the electrolytic generation of oxygen from water as a main source for atmospheric oxygen. Oxygen production would include the presence of electrolytically active oxidation–reduction interfaces at the water-saturated fringes in geological media, the deep-ocean oxygen concentrations (commonly 4 oxygen) far below the influences of the photic zone, and the high oxygen demand below the deep scattering layers.

1.3   BACTERIA: HUMAN PERSPECTIVES

The general public views bacteria in a very negative manner because of the sweeping pandemics of the past that wiped out large fractions of the population. Examples abound—from black death to typhoid, dysentery, cholera, tuberculosis, and botulism. Improvements in public health services and the appropriate use of vaccinations brought these pandemics under control. Bacteria were in general considered to cause many diseases in the early 20th century and emphasis was placed on identifying the pathogen and then developing a suitable manageable control strategy. It was indeed fortunate that these major bacterial pathogens could be isolated, cultured, and identified using selective agar-based culture media. The unfortunate consequence was that most bacteriological research involved the use of agar media that restricted investigations to bacteria able to grow on agar and they represent the minority. Most current investigations continue using agar media despite the limitations. This means that scientific precision may be claimed but it applies only to a limited number of bacteria species.

It is not surprising that after the culturable bacterial pathogens were identified, research turned toward viral pathogens and intense efforts were directed to the isolation, identification, culture, and control of these pathogens. This reached such intensity that in the later years of the 20th century, the mass media constantly informed the public that all pathogens were viral, thus discounting the role of bacteria. The advent of *E. coli* strain 0157 again changed public perception and demonstrated that bacterial pathogens continue to be important. Public health facilities again recognize the risks to society associated with bacterial pathogens. Modern research now emphasizes reductionism—analyzing complex organisms into simple constituents. For example, bacterial identification reduces bacteria (as complex organisms) to significant enzymes, genetic components, and biochemical signals.

One serious problem with the science of bacteriology has been that covert introduction of reductionism led to the exclusion of most bacteria from the identification process. In the natural environment, it is common for a bacterial biomass not to have been composed of and developed by a single bacterial strain, but act as a collectivistic structure involving many bacterial strains. Activity of each strain within the collective biomass would then be controlled by the group of strains as a whole. In natural circumstances, the numbers of strains can vary with the nature of the environment. In highly polluted conditions, the number of recoverable strains may be as few as six to eight. This is because the pollutant(s) significantly impact the bacterial community (consorm), leading to the survival of strains that can adapt to and take advantage of the pollutant(s).

In a pristine natural environment, it is common for 30 to 60 bacterial strains to inhabit a single community. When the environment becomes stressed by some dominant factor (such as a pollutant or temperature), the consorm adjusts and the numbers of strains shrink to strains able to adapt to the new conditions.

The primary quest in bacteriology has been to locate clinically or economically significant strains using reductionist techniques that tend to exclude any strain that cannot be conveniently cultured on agar media regardless of the significance of that strain to the collective functioning of the consorm. When evaluating, culturing, or identifying a consorm that would commonly be an interacting mixture of 6 to 60 strains, the first challenge is to address the dynamic nature of the consortial biomass. Dynamic changes will be expected based on changes in the external environmental matrix and the internal shifts within the biomass as it grows and responds to these conditions. The approach to identifying these consormial forms of bacterial biomass should not be reductionist and should be from the view of a critical ecologist looking at the site environment and the response of the biomass to that environment. The response may take the form of structural or locational changes with the bacterial consorm adjusting accordingly.

A natural environment can be expected to contain many different bacterial strains that are inactive. This inactivity is common in the bacterial kingdom and can take the form of resting ultramicrocells, endospores, starved cells, or cells protected within extracellular polymeric substances by tightly bound water. In these cases, the bacterial community that forms cooperatively within the active consormial biomass will most likely be reflective of bacteria that adapted to that cultural habitat.

In some ways this approach to bacteriology practiced by "lumpers" (using a collective communal approach) and "splitters" (using precision endowed by reductionist techniques) has created two pathways for the identification of bacteriologically influenced events. When there is clearly the need for precision at the strain level dictated by the need to precisely identify a pathogen, the reductionist approach would appear more appropriate. However, that is only the case if the pathogen can be monocultured. If it cannot be cultured using currently available techniques, it may be ignored and the cause of the disease would fall to another source. It is possible that some cancers are in fact triggered by bacterial pathogens that have not yet been isolated or identified in vivo. In the lumper approach, precision becomes more challenging since the composition of a bacterial community can shift as the biomass matures and the environment responds to the growth.

This atlas specifically addresses the potential to identify bacterial communities using the lumping approach. The classification is more pragmatic and focuses on the practical consequences of findings without attempting to achieve the type of precision that would be essential for the clinical diagnosis of a bacteriologically influenced disease. Application of the atlas is designed more to provide a means to classify bacterial communities (consorms) in environments where these growths are significant. In industry, the most significant bacteriologically influenced event is corrosion, particularly of steels and concretes. Corrosion is also an important factor in water production and quality control. Both activities may be impacted by bacterial consorms.

The oil and gas industries must determine not only the role that bacteria play in the generation of their products, but also the roles they play in plugging wells and pipelines, leading to premature well field failures. In agriculture, emphasis has been placed upon the role of the Rhizobium species in fixing nitrogen in leguminous crops but not upon the complex bacteriologically driven consortial activities that occur routinely in soils. Soils have the additional challenge of the location of the oxidation–reduction interface near the surface at the static water level. This creates a set of lateral interactive consorms that also interface with the rhizosphere and directly with plant roots.

Meteorologists need to understand and accept the nature of bacteriological activities in clouds from the perspectives of controlling precipitation and acting as bioaccumulator and biodegrader for chemicals emanating from the surface as rising particles or volatiles. Ice formations can also involve bacteriological activity not only at the higher temperature (2 to 7°C) required for formation of ice, but also the ability of more bacteria to grow in ice at temperatures as low as −18°C. Large populations of bacteria can also be recovered from gas hydrates that maintain high methane:water ice ratios and grow primarily on the continental shelves around the planet. These gas hydrates form some of the largest energy reserves of natural gas (methane) on the planet, exceeding the combined energy reserves for conventional gas, oil, and coal by a factor of two.

This atlas attempts to adhere strictly to the lumper concept that involves dealing with bacterial communities within a biomass and the environmental conditions within which these growths occurred. Classification is based primarily on the in vivo environment with emphasis on the ORP and the form of the water environment within which the consorm is growing. In this approach, individual strains of bacteria are considered subservient to the consorm and will only remain significant components if they contribute to the functioning of the consorm.

Culture of a bacterial consorm is dependent upon the creation of environmental conditions observed at the original site of observation. However, complete duplication of the conditions may not be required; only those factors critical to the activities of the bacteria within the consorm may be needed. Some level of flexibility, particularly to ambient pressures, must be recognized because of the considerable tolerance of bacteria to high hydrostatic pressures and their limited ability to function at different temperatures. Critical factors that can dominate the ability to culture consorms include (1) the ORP at the desired site of culture, (2) suitable surfaces with appropriate porosities, (3) supportive nutrient regimes, and (4) an inoculum volume ranging from 2% minimum to 10% maximum of the acquirable volume within the cosm in which the bacterial consorm is to be grown.

2 Common Bacteriologically Initiated Events

2.1 HISTORICAL OVERVIEW

Science has quite naturally become preoccupied with all matters that directly impact the human species. As a result, it is not surprising that the greatest attention in bacteriology has focused on identifying and managing to control the recognized bacterial pathogens of humans and economically important animals and plants. During the 19th century, two major events affected the direction of bacteriology as a science. The first was the political need to combat bacteriologically influenced diseases, particularly when populations were becoming concentrated into industrialized communities. The second was the uncertainty whether bacteriology was a subdiscipline of zoology or botany or, from a religious perspective, simply the work of the devil. Louis Pasteur resolved the latter concern with his elegantly simple proof that there was no such thing as spontaneous generation by evil forces such as Satan; the devil was not involved.

Throughout the 19th century, botanists and zoologists continued to conduct turf warfare over who should control the new and potentially rewarding kingdom of bacteria! One outcome was an agreement that the classification system for bacteria should follow the classical Linnaean forms. This would mean that all bacteria would be divided into genera and species, each of which would be unique and different from all other genera and species. This system worked well for plants and animals that essentially pursue their lifestyles independent of all other plants and animals. For the bacterial kingdom, this system did not work so well, simply because bacteria function and behave differently.

Unlike plants and animals, bacteria commonly function as communities composed of many different types that work together to allow efficient functioning of the community. No single strain dominates continuously within the bacterial community. Unlike plants and animals, the community is dynamically driven by the various strains of bacteria influenced by shifting conditions within the immediate environment that encapsulate the vibrant community. This shifting environment drives changes in the bacterial strains active within a community. This dynamism has been largely rejected as a significant factor in classical bacteriology. During the past two centuries, the natural scientific approach has been to "divide and conquer." In other words, separate all the bacterial strains, grow the strains individually, and identify them independently as unique organisms (like plants and animals). This creates the need to further separate their components for independent identification but this caused two major problems: (1) many strains were lost through because they were not culturable using the methods applied; and (2) the ability to identify a bacterial community as a prime functioning living entity was lost because emphasis was placed on the culturable survivors.

In the past 200 hundred years, bacterial communities have been neglected and the terminology applied has been limited to basic descriptors: slimes, ochres, plugs, biofilms, nodules, tubercles, encrustations, pits, perforations, and stinks. In the applied world dominated by science and engineering it has been relatively convenient to reduce the activities of these bacterial communities to the relatively insignificant or consider their activities as unimportant factors in modelling processes. In reality, these descriptors caused a devaluation of the importance of these organisms (for example, slimes are nuisance growths over surfaces that should be removed; ochres are geochemical and lack bacteriological activity; plugs are physical clogs that have no bacteriological components).

Implications in science, medicine and engineering abound from the premise that bacteria function in community structures rather than at species level and should be identified as dynamic and ever-changing communities. To date, progress in bacteriology has occurred primarily by splitting communities only into groups that are culturable and identifiable as separate entities instead of grouping all the bacteria within a community and identifying the community in vivo. This approach presents a number of weaknesses:

1. The role of a species within the community is only as strong as the uniqueness of the contribution of the species to the community. If another species takes on the event that other species take over the unique roles, the initial species may be eliminated.
2. The nature of the bacterial cell (simple form and function) means that it is relatively easy for genes from another species (or a plant or animal infested within the community's environment) to move from the cells of one component species to the cells of another bacterial species within the community of interest. This would render the characteristics of the community as more reactive and dynamic.
3. The nature of a bacterial community imparts a greater ability to adapt to changing environments than a single species has. Thus, the order of dominance within the community will change, even to the point that some species are eliminated and others achieve dominance.

Classical Linnaean categorization has been applied to the bacterial kingdom primarily by reductionists. This approach does not allow recognition of communal adaptation as a state of constant dynamic change in the dominant bacterial species within a continuously changing community structure.

This atlas therefore will not use the classical systematics developed primarily to identify pathogenic species. It will use a more pragmatic approach invoking the principles of Occam's razor (the objective is to use the fewest number of possible assumptions to explain a thing). For the bacterial kingdom (the "thing"), the first assumption is that most bacteria within a natural environment function in community structures and not separately as independent species. The second assumption is that these intelligent communities are driven not by their species composition, but the source of energy drives the community. The third assumption is that the nature of the community can be best determined by its structure, form, and function of the community within its environment. These assumptions when applied to the classification of bacteria mean that the first step of identification must be the recognition and determination of the communities.

History provides a record of events recognized either through the literature or practices of societies throughout time. Bacteriology, as practiced today, follows many assumptions that have come down through history as scientifically valid. The history of the science reveals a series of events that created the present mindset. The first of the major events that shaped bacteriology was the development by Anton van Leeuwenhoek of a single-lens microscope to describe microorganisms including bacteria active in many different environments. His diligent observations were documented and have stood the test of time as some of the first observations of microscopic organisms. So extreme were his talents for using the single-lens microscope that no one subsequently succeeded in mastering the instrument to that extent. Not until the development of the compound-lens microscope were Leeuwenhoek's observations of bacteriological activities surpassed.

The next major event was finding a bacterial culture method. Plants and animals offered researchers the convenience of growing in correct environments. By comparison, bacteria were far more challenging because they were ubiquitous in all the environments explored with the compound microscope. Some activities such as iron pans, ochres, corrosive pitting of cast irons and steels, pig iron, and tubercles were considered purely geochemical in origin. As of 2009, the true extents of bacteriological influences on these primarily geochemical artifacts remain to be fully developed. Since 1880, the predisposition to culture bacteria as defined culturable strains and/ or species rather than as communities has grown. This led to the use of sterile potato surfaces upon which the bacteria could be grown as identifiable colonies. The next development was the use of gelatine surfaces to grow many disease-producing bacteria; the disadvantage was that the gelatine could be degraded. As a result, agar-gelled surfaces appeared in the 1880s.

Today the agar plate has become the normal medium for bacteriological culture techniques, primarily because the gelled surface allows the development of colonies via the growth of a few cells inoculated onto the surface. The advent of agar for selective culture led to the development of methods to identify many bacterial pathogens, e.g., the causes of anthrax, cholera, and typhoid. The precision of agar culture techniques allowed identification of selected bacterial cultures.

2.2 CHALLENGES OF CLASSIFYING "UNCULTURABLES"

One challenge of selective use of agar-based culture media is the limited, albeit ill-defined, range of bacteria that may be cultured. By and large, bacteria that cannot tolerate and be cultured on agar remain neglected areas of the science of bacteriology. Fortunately, many major bacterial pathogens of humans and economically important animals can be cultured on agar and identified from agar-based cultures. This reliance upon agar persists and the inevitable consequence is that bacteria that cannot grow on agar are ignored. Recent advances allow agar-cultured bacterial strains to be subjected to extensive biochemical identification systems based on nucleic acid configurations, selective fermentation functions, or the methyl ester fatty acid compositions of cell walls. Despite these advances, any bacteria unable to grow on agar-based culture media are not recognized as being identifiable.

A conceptual divide exists between bacteriologists who "split" identification down to the molecular biochemical level and those who "lump" bacteria into definable communities. Splitters want to define all parameters with maximum biochemical precision—to the point of excluding bacteria that are unable to grow or remain active through selective cultural processes. This means that splitters who are preoccupied with maximizing precision avoid working with bacterial cultures that cannot be readily grown using common agar-based selective cultural practices. Lumpers, in contrast to splitters, are preoccupied with a bigger picture: culturing, identifying, and conducting research on bacterial communities.

The nature of these communities that involves dynamic interactions between the component bacteria within the community and also with neighboring communities allows integration within a common environment to form a consorm. Scientific investigation of a community or consorm creates additional challenges in pursuing precision because of the continuing potential for the activities of a community or consorm to change via realignment of the component bacterial strains. Most bacteriological progress is being made by the splitters, buoyed by the need for precise identification, particularly of pathogenic bacteria. By contrast, advances of bacteriologists who support the lumping approach remain limited because many community structures remain to be reliably identified and their activities quantifiable in a realistic and precise manner.

Medicine also has a continuing need to rapidly identify specific bacterial pathogens. The splitters provide the mechanisms to achieve this goal. In the environmental and engineering fields, the need is to develop the science of "lumping" identifiable communities or consorms as bacterial groups together. The primary goal of this atlas is to develop a classification system suitable for the identification of bacterial communities relevant to various scientific disciplines.

2.3 EVOLUTIONARY TRENDS TOWARD BACTERIAL DIVERSITY

One important goal is to learn from the past and determine how it will affect the future. Unlike plants and animals, bacteria have left no observable record of their evolution. Animals displayed dramatic evolution through the dinosaur period. Plants have evidenced an evolutionary shift from seaweeds to ferns to trees. No strong observable evidence exists fir bacteria. The prime focus—the manner in which the earth evolved to become able to sustain life—is more in the realm of speculation than a sustainable scientific hypothesis. General conjectures can be made about the evolution of favorable environments for the development of bacteria. It is generally surmised that Earth was formed as a planet around four billion years ago. Since then a number of critical environmental events have occurred:

1. Earth cooled as a result of a reductive atmosphere.
2. Water collected in clouds orbiting the Earth in a manner similar to those clouds seen around the planet Venus today.

3. Earth cooled further, allowing water to impact on its hot crust.
4. Earth cooled further; condensing liquid water could penetrate its crust.
5. Vigorous water cycling formed in the atmosphere from formations of groundwater in the crust and pooling surface waters.
6. Liquid water moved through the crust, reacting and causing swelling due to water sorption and steam formation; Earth's anhydrous state shifted to a hydrous state.
7. Electrolytic and geochemical reactions with permeating water caused fundamental changes in the atmosphere from a reductive to an oxidative form as electrolysis of the captured liquid water separated into oxygen and hydrogen.
8. Earth cooled and swelled with the admission of water into the crust; water is also pooled into depressed surfaces above water-saturated crustal elements.
9. Clouds formed from global cooling; sunlight reached the pooled water collecting into seas, lakes, and rivers to trigger a water cycle.
10. Complex chemical reactions in the liquid groundwaters and surface waters generated reproducible forms of organic biomass.
11. During this evolutionary period, Earth was continuously bombarded with ice and rock meteorites, presenting ongoing potential for inoculation of pooling waters with life forms from the ice and rock.
12. The atmosphere became oxidatively stable due to adequate electrolytic generation of oxygen; this process was later supplemented by photosynthesis.
13. The broad spectrum of life in the form of microorganisms, plants, and animals evolved to the range observed in modern environments.

2.4 TWO-DIMENSIONAL GRID DEFINITION OF BACTERIAL COMMUNITIES

Debate continues about the time scale of these 13 stages of evolution on Earth. Extreme views range from as short as four thousand years to as long as four billion years, but these estimates involve a considerable level of conjecture. Bacteriological activities on Earth may be considered to have started during stages 6 and 7 with the condensation of liquid water along with the potential for extraterrestrial inoculation from incoming ice-rich debris and the intrinsic evolution of processes essential to life forms in vivo.

The challenge now is to develop a methodology to identify bacterial communities of functional significance rather than following the Linnaean classification that divides down to levels of genera, species, strains, and taxons. Since the birth of bacteriology, the splitters have ruled, based on clear successes in identifying many significant bacterial pathogens and biochemical validation at the molecular level. Lumpers have not achieved significant credible successes to match those of the splitters!

This atlas is designed to place all bacterial communities into a convenient two-dimensional grid to allow their positioning and recognition. Such positioning should consider the primary factors that influenced the evolution of bacterial communities on

Earth. Based on speculation about the prime functions involved in evolution on Earth, these primary factors are (1) the oxidation–reduction potential (ORP) and the physical state of water (PSW) expressed as viscosity. In this approach to the classification of bacterial consorms, the defining grid is composed of the ORP range within which a particular consorm is active as the horizontal axis (latitude) and the PSW conducive to the activity of the consorm is the vertical axis (longitude). The intrinsic ORP and PSW are critical to the types of bacterial consormial activity within a given environment.

ORP is an intrinsic aspect of saturated crustal groundwaters and to a lesser extent is also important to pooled surface waters, primarily because ORP is a reflection of ongoing geo-electro-magnetic function in an environment saturated by water. ORP is commonly measured in waters as ranging from –600 to +200 mv. This electrolytic spectrum indicates that the negative voltages tend to cathodically drive water to form hydrogen while the positive anodic voltages drive the water toward the formation of oxygen. There are some clear differentiations of bacterial communities on the basis of whether they are anaerobic and live in reductive environments (negative ORP values) or are aerobic and live in oxidative environments (positive ORP values).

The transitional environment in which ORP values are mildly oxidative or reductive constitutes a reductive–oxidative (redox) front where biomass tends to congregate in groundwater situations. Usually redox fronts are generated as the environment transitions from reductive to oxidative, with ORP values ranging from –50 to +20 mv. Within environments in which groundwaters move from a reductive to an oxidative state, the oxidative–reductive interface provides a unique site at which nutrients flow inward from the reductive zone and oxygen flows in from the oxidative zone. Thus, this interface provides a suitable site for the aerobic degradation of organics that were recalcitrant under reductive conditions. The growth of biomass at the redox front is therefore stimulated by the movements of organic nutrients from the reductive zone into a region where oxidative degradation can occur rapidly.

Bacteria are affected by the ORP level in the environment that appears to impact aspects of community survival, growth, and reproduction. ORP also exerts major impacts on the types of metabolisms that develop. In practice, ORP can be categorized into four major groups based upon the types of communities that are likely to dominate:

- R – Very reductive environments with ORP values ranging between –600 and –200 mv
- G – Moderately reductive conditions in which the ORP commonly ranges from –200 to –50 mv
- F – At the redox front (interface), oxidative and reductive conditions cause vacillations between ORPs of –50 and +20 mv
- O – Fundamentally oxidative conditions in which the ORP normally exceeds +20 mv

Within each of the four ORP groupings, challenges are created by the consortial biomass generating internal ORP gradients within the growths. This means that, for example, a consorm dominated peripherally by iron-oxidizing bacteria (IRB; ORP types F and O) may also contain communities that are commonly active in more reductive environments (types R and/or G). This raises the issue of using ORP as a primary feature in classifying bacterial communities because ORP is micromanaged within the generating biomass, particularly at the redox front (F) where subgroups may interact with G and O groups. It is therefore possible only to broadly categorize consorms based on the specific environments they dominate and take into account the fact that ORP values within a biomass can vary in a relatively stable manner.

The ORP gradient within the crust, surface waters and groundwaters may appear primarily oxidative because free oxygen is available through interaction with the oxygen-rich atmosphere and mainly reductive (O to R) at greater depths away from the influence of atmospheric oxygen. The question remains whether the geo-magnetic forces permeating the crust can cause electrolysis of the ambient water, leading to the generation of oxygen and hydrogen. Such an event would mean that oxygen could be created within very reductive environments where water is present. Whether this electrolytic function is purely physico-chemical or involves bacteriological interference remains conjectural; the potential exists for bacteriologically driven oxidative activities deep within a reductive (R) environment.

Functionally, the best symbol for the ORP gradient is determined by the status of the carbon in the immediate environment. In the atmosphere, the dominant gas naturally present is carbon dioxide (CO_2) as a daughter product of the oxidative degradation of complex organic forms driven by the twin processes of respiration and combustion. Methane (CH_4) may also be present in the atmosphere but arises from vented natural gases coming from reductive environments; this gas is only degraded, not generated oxidatively. Other volatile organically based products may also occur in the atmosphere but the ORP gradient would primarily be established in the crust. The carbon dioxide at the oxidative end of the gradient would gradually be reduced to elemental (e.g., coal) or reduced (e.g., oil and gas) forms of carbon.

Transitional molecular forms of carbon from the most oxidative to the most reduced would incorporate (in oxidative to reductive environments, respectively): carbonate (CO_3), carbon dioxide (CO_2), carbon monoxide (CO), petroleum hydrocarbons (C_XH_Y), natural gases (CH_4), and then carbon (C). Essentially these different carbon forms relate to the ORP gradient grouping: group R dominated by petroleum hydrocarbons (C_XH_Y), natural gases (CH_4), and then carbon (C); group G dominated by natural gases (CH_4); group R dominated by carbonates (CO_3), carbon dioxide (CO_2), carbon monoxide (CO); and group O dominated by carbon dioxide (CO_2) and escaping natural gases (CH_4). These groups may be defined by their atomic C:H:O ratios. For the four defined groups, this atomic C:H:O ratio would be: (R) 1:>1:0; (G) 1:>2:0; (F) 1:<1:>1; and (O) 1:0:>1. This shift in the dominance of hydrogen (groups R and G) through to oxygen (group F) and then to the dominance of molecular oxygen (group O) can be linked to the dominant bacterial consorms present in these different ORP environments. The four ORP groups are also very distinctive in the types of bacteriological dominance observed.

On the basis of ORP-defined grouping of bacteria, each of the four ORP groups would have a unique consormial dominance. For example, group R is likely to be dominated by methane-producing bacteria; group G is more likely to be dominated by petroleum hydrocarbon-producing bacteria along with sulphate-reducing bacteria; group F would be a complex collection of aerobic and anaerobic bacteria functioning collectively; group O would be dominated by aerobic bacteria. These characteristics clearly show that the ORP clearly differentiates several major bacterial communities using the x axis (latitude) in the framework of an atlas based on community identification.

In the investigation of any bacteriological activity, the state of water as a cultural matrix becomes as important as ORP level. It is proposed that the viscosity of the water serves as the longitudinal (y axis) in the conceptual atlas format. Along with the environmental impact of ORP on the abilities of bacteria to become active as precursor of growth; the availability of water provides a responsive matrix within which bacteria can function. Furthermore, the characteristics of the water are directly impacted by these bacterial activities. Water forms the bulk of a cell in every living organism and also provides capsules around cells for many bacteria and some aquatic animals and plants. In bacterial communities, unlike plants and animals, the water outside cells often provides matrices essential to activities and survival. Many bacteria extend their influence to the water outside cells (extracellular water) almost as if the cells control the water. In plants and animals, cell function concentrates in the water inside the cells (intracellular water) through complex structures. Bacteria have structural components that help to manage intracellular water. Essentially, bacterial cells exhibit far greater ability to control extracellular water, primarily via releases of extracellular polymers (string-like water-absorbent molecules).

In the classification of bacterial communities, the role and form of extracellular water define the type of bacterial activity. Since the primary function of extracellular water is support for bacterial cell functions, its properties are critical to the defining the form of bacterial community that develops. This support, generated and manipulated by bacterial, cells leads to a number of functions categorized as accumulative and protective.

Accumulative functions generated primarily by extracellular polymeric substances are created by the structuring of the water into a more viscid state. Since a bacterial cell is both small and relatively structureless, the polymeric structures outside the walls around cells provide means to compartmentalize and accumulate specific chemicals that may have some eventual value to the cells. Essentially this extracellular water develops a form that can fit tightly around a cell in the form of a capsule or more often a diffuse envelope whose volume may be many times that of the cell itself. Generally this accumulative function causes the concentration of organic molecules, nutrients, and metallic cations. All of these accumulations are initially locked within the polymeric structures in the water for eventual use by bacteria forming a capsule or envelope of entrained water.

Organic molecules such as natural gases, volatiles, petroleum hydrocarbons, and daughter products of degradation held within polymeric structures are subject to assimilation outside of or within cells by oxidative or reductive biochemistry. Under reductive conditions, the end daughter products are most likely to be shorter chained fatty acids, petroleum hydrocarbons, and natural gases. Under oxidative conditions in which degradation is more complete, the likely major terminal product would be carbon dioxide. Volatile organics tend to become entrapped within the bound water inside the polymeric matrix and (1) restrict diffusion and (2) encourage degradation under oxidative conditions. Under such conditions it would also be expected that the petroleum hydrocarbons would degrade and serve as sources of carbon and energy for ongoing bacterial activities.

In the bioaccumulation of molecules within the polymeric weave of bound water held in the capsule or envelope, many cationic metals such as ferric and manganic forms as oxides and hydroxides accumulate. These concentrations far exceed the metabolic requirements of the bacterial communities and can continue to bioaccumulate around the cells in the capsules or envelopes. They produce secondary effect, particularly under oxidative conditions, leading to the formation of encrustations. Where iron dominates (in various ferric forms), the whole growth begins to harden and iron concentration rises as high as 85 to 95%. The growth resembles rusticles, ochres, pig iron, or iron pans and shifts from a bacteriologically dominated structure to a geo-chemical form exerting little observable bacteriological activity.

In addition to the bound water held within a capsule or envelope by polymerics acting as accumulative storage vesicles, the accumulated materials also provide protective functions of two primary types. First, the permeability of water through the capsule or envelope reduces as the metal content rises and the porosity falls correspondingly. Second, metals concentrating in the biomass gradually become inhibitory and then toxic to potential predators (particularly protozoa, insects, and fungi). Generally, as the ferric content of the growing biomass rises from trace amounts to 20%, the growth no longer serves as suitable source of nutrients for predators. Another potential protective function relates to the generation of antibiotics that reduce the potential for invasion of the biomass by other bacteria.

The growth of a biomass with high chemical content involves another significant protective function related to the growth of at least some bacterial communities. The function is the formation of a cloistered region of growth where activities are surrounded and protected by a high-density outer layer. Commonly this cloistered growth is more likely to become reductive and dominated by anaerobic bacteria held within a complex biofilm (slime). Several common forms of iron-rich growth offer this type of protection to bacterial communities within a biomass: nodules, tubercles, and encrustations. Ochres often display these types of terminal growths. A nodule has an even protective coating commonly shaped like a mountain with a rounded top. Tubercles tend to appear more ridge-like in form and their tops are usually irregular with multiple dome-like structures. Encrustations are more irregular in form and are usually attached to and remain near flatter surfaces.

These iron-rich growths can generate cores of intense bacterial activity that may be dominated by hydrogen sulfide production, causing electrolytic pitting and perforations when attached to steel surfaces. This phenomenon is known as pitting perforation corrosion in the steel industry and commonly is related to sulphate-reducing bacteria. With some encrustations and ochres, the core of intense bacterial activity often is dominated by acid-producing bacteria that cause the pH at the attachment surfaces to drop into the acid range (commonly, 3.5 to 5.5) as a result of fermentative processes producing short chain fatty acids. When this occurs on steel, its strength gradually reduces and its porosity increases. Pitting followed by perforation will cause steel to lose strength and fail, possibly with weeping of water through the pipe walls. These activities all involve some protective functions in the biomass that allow core bacterial communities to survive and flourish.

As a biomass increases in size and hardens due to chemical accumulations within the capsules or envelopes around the cells within the communities, a series of secondary effects impact the surrounding environment. These effects are created when the biomass occludes (blocks, plugs, clogs) the passage of water through or around it. In groundwater situations, such growths can actually cause water to stop flowing through the impacted groundwater site (e.g., a plugged water well).

It is probable that water flowing through an aquatic or groundwater environment will be impacted by the polymeric structures generated by the bacterial communities it contains. These polymers affect the viscosity of the water. It is proposed that viscosity serves as much as the ORP as a fundamental influence on the types of bacterial communities that thrive in given environments. However, a primary difference is that ORP defines the environment within which the organisms can be effectively active. Water viscosity is a state affected by the organisms during their periods of activity within that environment. Effectively the x and y axes are driven by factors influencing the ability of the organisms to be active (ORP, latitude, x axis) and the factors that are products of organism activities (water viscosity, longitude, y axis).

Viscosity (physical state of water or PSW) is a function of the properties of organisms and chemistries held within or interfacing as a surface with the water. The viscosity of the water is measured generally in centipoise (cP) and pure water has been found to have a viscosity of 0.894 cP at 25°C. Viscosity increases with lower temperatures and water at 20°C has a viscosity of 1.002 cP. In natural water-based products, viscosity is clearly a reflection of chemical additives.

Examples of this PSW include honey (2,000 to 10,000 cP), chocolate syrup (10,000 to 25,000 cP), and tomato ketchup (50,000 to 100,000 cP). Under extreme conditions, pitch, a compound high in petroleum hydrocarbon content, has a viscosity of 2.3×10^{11} cP.

The viscosity of water in living systems is impacted by the relationship of molecular water to adjacent charged organic and inorganic surfaces. Thus, water viscosity is impacted on micro and even nano scales, rendering measurement (commonly in cP) very challenging. In generating the characterization of bacterial communities through their impacts on water viscosity, a considerable level of speculation must be based upon the observation of the manner in which bacterial communities react and control the water within their environmental domains.

2.5 ESTABLISHMENT OF GRID LOCATION POINTS FOR BACTERIAL ATLAS

The atlas approach proposes to use water viscosity (in cP) as a prime differentiator of the different bacterial communities as the y axis (longitude). Observations of how bacterial communities operate and develop within consorms make it possible to use the viscosity of water as a prime differentiator based on the premise that pure water has a viscosity of 1.0 cP. It is reasonable to assume that the viscosities of all waters intervened in some manner by bacteriological activities or products of such activities will commonly exceed the base value by one to five orders of magnitude.

Extreme circumstances in which water is reduced to a minor fraction of the bulk content will lead to extreme viscosity. For example, pitch as a potential bacteriological product with very high petroleum hydrocarbon content would commonly exhibit viscosities of eleven orders of magnitude (log 4 cP). For the purposes of this atlas, the water viscosity range is on the logarithmic scale from zero to log four (the equivalent of 10,000 cP). Separation of the y axis would utilize the log cP values where <1 would indicate water viscosity below 1, possibly because the water is dispersed into a gaseous or aerosol form (0 viscosity). Values between log 1 and 2 cP reflect viscosities observed in non-colloidal free water. Values between log 2 and 3 cP indicate that the water is viscid and bound, most probably by bacterial polymers.

After water becomes more bound and encapsulated within chemical matrices, its viscosity will rise into the range of log 4 and higher. At higher log cP values, the water becomes irreversibly bound within chemical matrices. A good example is water entrapped and bound within clay-based matrices that cannot be removed even by heating well above 100°C!

2.6 SUMMARY OF BACTERIAL COMMUNITY GRID POSITIONING ATLAS PRINCIPLES

The basic nature of this bacteriological atlas proposes that the x axis be established through the ORP expressed in millivolts (mv) present as the electrical potential within the environment. This sets the stage for the locations of various aerobic and anaerobic bacteria by their environmental positions along the oxidation and reduction gradients. The four divisions are R, reductive; G, gradient to reductive; F, front where oxidative and reductive conditions interface; and O, oxidative. The R division encompasses very reductive environments with ORP values ranging from −600 and −200 mv. The G division conditions are still reductive but only moderately so because the ORP value commonly ranges from −200 to −50 mv. Most bacteriological activity occurs in F division at the redox front—the interface between reductive and oxidative conditions that commonly causes environmental fluctuations of ORP values between −50 and +20 mv. O division occurs when fundamentally oxidative conditions of ORP normally exceed +20 mv. Thus the x axis allows primary differentiation of bacteria into the R, G, F, and O divisions ranging from fermentative (anaerobic) to predominantly oxidative (aerobic with respiration). The x axis therefore dictates the environmental ORP conditions under which various bacterial communities will thrive.

The y axis reflects the impact of bacteriological activities on the viscosity of the water. Grid analyses for the atlas are therefore separated into four primary groups (R, G, F, and O, from reductive to oxidative, respectively) for the x axis and five primary groups (0 or <1, 1, 2, 3, and 4, from very low to very high viscosities when measured as log cP values). The primary atlas grid contains 4 × 5 or 20 individual squares. Each of the 20 squares is divided into a 6 × 6 internal grid to allow differentiation of the defined bacterial communities. The atlas grid now contains 20 × 36 or 720 total second-level grid squares. This means the atlas can incorporate up to a maximum of 720 bacterial consorms (communities), assuming that only one consorm is placed within each second-level grid square.

Each square would be locatable using a simple numeric sequence by which all rows and columns are numbered starting with 1-1 at the upper left hand corner as. Each of the 720 squares could be found by a two-numbered locator for the column, followed by a hyphen, then a two-numbered locator for the row. For example, 10–12 would refer to the tenth column (in the G range) and the twelfth row (in the 1 range) and would be used to locate the bacterial consorm positioned at that site. The combination of two numbers and hyphen followed by two numbers (XX-YY) is called a focal point locator and represents the fmv (XX, x focal point based on ORP millivolts) followed by fcP (YY, y focal point based on water viscosity in cP).

This atlas is based on the premise that each bacterial consorm has an optimal requirement for both ORP (oxidation–reduction potential) and water viscosity, with the aerobes favoring the former and the anaerobes the latter. Bacterial consorms are also affected by the form of available water defined by the viscosity of the water within the environment expressed in log cP. While a particular grid square may be allocated as an optimal site for the activities of a specific group of bacteria, some variation in the range of ORP and viscosity allows a bacterial consorm to function. The individual secondary grid square therefore serves as a focal point to locate a particular consorm; no two consorms can be located within the same focal point. This variation is addressed in the atlas and as is considerable overlapping of the various groups.

Each bacterial consorm is defined within an alpha group of similar groups of bacteria. These consorms are defined primarily in Chapter 4 and covered in detail in Chapter 7. In the focal XX-YY location positioning (fmv-fcP), the XX potion reflects the ORP where the bacterial community is active. A smaller number would indicate more reductive conditions (01 [extremely reductive] as the XX value. An XX value of 24 would indicate an extremely oxidative supportive environment. Numbers for the fcP component range from 1 to 30 for the YY value; 1 reflects water dispersed into an aerosol diffused into the atmosphere and 30 indicates water bound into a very viscid form. Conversion of these fmv and fcP values forms the basis for the gridded atlas detailed in Chapter 10 and summarized in Table 10.17.

3 Bacteria Are Everywhere
Classification of Alpha Groups of Bacterial Consorms

3.1 HISTORICAL OVERVIEW

The human eye tells us what we can recognize. It was easy in the past to recognize plants and animals. Plants did not move about except when growing in water. Animals moved about a lot, whether in water, on the ground, or even in the air. From the beginning of life on earth, the eye has served as the prime detector of life by observation of movement. This reliance on sight reached the point where anything that could not be seen was considered not living! Until 200 years ago, visibility was the basic premise upon which life was recognized: if you could not see it, it could not be alive.

In the 17th century, Anton van Leeuwenhoek invented the first simple one-lens microscope and his eyes were the first to see microbes. He noted that they grew in all kinds of places and wrote about his findings, including descriptions of bacteria and other microbes. His work was widely followed. Unfortunately, the skills required to operate his simple one-lens microscope using candle light for illumination were beyond the abilities of most who wanted to duplicate his findings. Thus his findings were read, but no improvements appeared until the improved design of the compound microscope in the 19th century allowed better observation of microorganisms. More researchers could conveniently observe the submicroscopic world and report their findings. This created enormous controversy. Religions categorized these small forms of life as works of the devil, and chemists thought such microscopic growths resulted from complex series of chemical reactions. The medical profession tended to favor the chemically based doctrine and described the microbes as chemical "miasmas." Zoologists thought all small life forms were animals and botanists claimed that submicroscopic forms of life were plants. These differing views created a battle for jurisdictional control that lasted through the 19th century.

The only common feature in the controversy was that all parties recognized the potential for microbial threats to be real. Some wanted to control the exploitation of the development. Four major professional groups (religion, chemists, zoologists, and botanists) all fought for or against the recognition (and control) of these subvisible living organisms. Reactions ranged from denial (religious groups) to reductionism (chemists who wanted to believe that all subvisible organisms were inanimate and not alive) to political dominance (botanists considered microbes as plants and zoologists considered them animals). This four-way tug of war continued through the 19th century and in effect suppressed microbiology as a discipline. Only in the last quarter of that century did microbiology finally emerge as a recognized new scientific field.

Essentially the hammer blow was delivered to the religious dogmatists by Louis Pasteur when he finally laid to rest the theory of spontaneous generation by eloquently showing that bacteria are everywhere and cannot be grown "out of thin air." The theory of "thin air" arose from the claims that all microbes were the works of the devil and they appeared spontaneously in muck, mud, and mayhem. Pasteur showed that microbes were indeed living organisms that could be controlled by heat. They could not be generated spontaneously in even the smelliest and most revolting muds after the application of heat. Religious objections were laid to rest and microbes were no longer considered works of the devil.

Robert Koch introduced scientific method to the new discipline of microbiology and it became possible to selectively grow bacteria in pure cultures. That discovery ushered in the golden age of bacteriology. Between 1875 and 1900, many major bacterial pathogens were discovered using techniques initially developed by Koch and his research followers.

Botanists and zoologists continued to argue whether microbes were plants or animals and a divide was reached in which fungi were considered plants and bacteria and protozoa were considered animals. This reduced the friction between the two groups and allowed microbiology to develop unfettered by infighting. One classical tradition originating from zoology and botany and still practiced in microbiology is the adoption of the Linnaean system of classification from kingdom down to genus and species.

However, unlike plants and animals, bacteria tend to function in communities involving many species and genera and share their genetic material (something animals and plants may or may not do). As bacteriology grew as a discipline, the need to break bacteria down into culturable units containing component cells that were exactly the same was universally accepted. The new challenge was growing these "pure" cultures of bacteria in a robust and reproducible manner.

A revelation during lunch in the Hess household in 1882 set the course of bacteriology into the 21st century! Frau Hess suggested to her bacteriologist husband that he should use a gelled dessert to grow his bacteria. He had been complaining that bacteria did not grow well on sterile potato slices and caused the gelatin-based gels to break down when he tried to grow discrete colonies of bacteria. She suggested using an agar-agar extract from seaweeds that was commonly added to jellies and it worked. Her simple suggestion opened a floodgate of bacteriological discoveries that continued through the end of the 19th century. By that time, cholera, typhoid, anthrax, and many other bacterial diseases were definable and better understood. Agar-agar then served as the preferred method for culturing bacteria. Its use dominates today despite vigorous attempts to use alternative gels such as those based on silica.

Agar-agar is an extract of seaweeds that is subject to changes in characteristics from batch to batch. What makes it so durable is that it resists attacks by bacteria that would otherwise cause the gel to collapse. This means that only very rarely do agar-agar gelled plates break down because of bacterial action. Today this durability and ability to culture major bacterial pathogens has placed agar-agar as the preferred culture for bacteria of interest. Few practitioners seriously consider the times when agar-agar failed.

Failures can be precipitated by: (1) inability of bacteria to mine water from the agar-agar (essentially a tug of war between agar-agar and bacteria); (2) chemicals in the agar-agar may prevent growth of the bacteria; (3) the surface of the agar-agar may provide too much oxygen to allow the bacteria to grow; and (4) the relatively thin layer of agar-agar may not constitute an environment favorable to bacterial growth. Fortunately, for medical bacteriologists, most bacterial pathogens were able to grow on agar-agar. Now, more than a century later, agar-agar culture media stand supreme and little consideration is given to the many bacterial species and communities that cannot grow on them. Thus, only bacteria that can be cultured in a definable manner are subjected to refined identification procedures.

Chemists have long opposed bacteria. They often view bacteria as little more than bags of chemicals (enzymes and nucleic acids). After the recognition of bacteria as distinct forms of life, chemists concentrated on methods to kill the bacterial pathogens without harming humans suffering from diseases. Paul Ehrlich, in the early 20th century, proposed combinations of organic chemicals with lethal metallic cations such as mercury and arsenic to differentially kill pathogens. Fame and fortune came with compound 606, which was found to be effective against syphilis—a three-phased, sexually transmitted, bacterial disease rampant in society at that time. The next antibacterial advance followed in the 1930s with the discovery of sulfonamides that exerted a broad spectrum of activity and were effective against many diseases. This "new age" specifically targeted chemicals that could be used to control bacterial pathogens.

At the same time as the first successful chemicals were found to "cure" bacterially influenced diseases, the ability of chemicals to both prevent and control disease was recognized. Between 1910 and 1940, it became established that bacteria-causing disease contained specific chemical components (antigens) that specifically aided infective processes. Another finding was that the human body generates neutralizing chemicals (antibodies) that blanket antigen activities. The science of immunology incorporated these findings into the identification of bacteria by their antigenic structures to develop vaccines and a new wave of methods to control bacterial diseases was established.

Agar-agar plates were studied widely since Herr Hess's discovery. One common observation was that some microbial colony forms could wage war on others. This inhibition activity was seen as a clearing circle under the growth that could destroy much of the growth within the spreading zone of influence. No one paid much attention to this detail until Alexander Fleming discovered a zone of influence in his cultures of Staphylococci originated from a fungal growth known as Penicillium. Fleming isolated and purified the zone and called the result penicillin. He observed one of the first effects of an antibiotic by discovering a microbe that could differentially attack and kill another microbe (i.e., Staphylococcus) without harming itself. Penicillin was clearly a very effective and selective toxic substance! Fleming refined the extraction methods for penicillin and by the early 1930s attempted to gain acceptance of the routine use of penicillin to treat bacteria infections. Unfortunately the

1930s were the "heyday" for sulfonamides. The pharmaceutical industry showed little interest in penicillin; it felt it already had the answer! However, sulfonamides became less effective because the targeted bacteria began finding ways to attack them. Fleming was persistent but it took a world war to trigger interest in antibiotics and 1940s and 1950s saw a surge in the production of diversified antibiotics.

Perhaps the next signal events that changed bacteriology were the discovery of deoxyribonucleic acid (DNA) and the initial attempts to crack the genetic code. After 30 years of effort, it became possible to understand parts of the genetic codes of some species of bacteria and then undertake sophisticated identification. Today the classification of bacteria is governed by the chemistries of their nucleic acids, the fatty acid structures of their cell walls, and their fermentative behaviors when presented with specific chemicals.

Investigations of all these systems require "pure" cultures of the bacteria grown preferably on agar plates. The bacterial world belongs to the splitters (bacteriologists who use biochemistry to define and identify specific strains). The caveat here is that the bacteria to be identified must be pure and commonly grown on an agar-agar substrate. If a bacterium cannot be grown on an agar-based culture medium or is not pure (growing within a community of several strains or species), sophisticated modern techniques lose value. The bacteriological world is ruled by splitters but in practical terms, bacteria generally are active in communities and this fact interests the lumpers (bacteriologists who believe that bacteria work in communities with other bacteria). Their mandate is to examine the collective bacterial intelligence of the whole community (consorm) rather than split conveniently culturable strains from the consormial web.

This book addresses the need for a bacterial classification system suitable for lumpers. To fill this need, it is important that the information relates more to activity levels and potential risks that the bacterial community may engender. Risks commonly relate to structural failures through microbiologically influenced corrosion (MIC) frequently involving bacteria and engineering failures caused by plugging of water, gas, and oil systems such as pipelines and wells.

One unique feature that sets bacteria apart from other microorganisms, plants, and animals is the fact that bacteria do not necessarily require oxygen for respiration. Plants and animals including protozoa and fungi all use oxygen in the absence of light (solar radiation). If liquid water is present, an organism always has the potential for activity and growth. However, bacteria as a consormial group, may be active in zones of the Earth far from any oxygen, for example, deep in water-saturated geological strata, in deep-ocean sediments, and even in oils that contain significant concentrations of water. This book will begin by dividing the bacterial kingdom on the basis of community structures associable with the water-laden regions of the Earth.

In summary, life on earth reflects an organic carbon-dominated biomass that exerts tremendous environmental flexibility and impacts virtually all sites where liquid water in some form is found under a wide range of oxidative and reductive conditions. Bacteria also require suitable forms of organic forms of carbon. All life as we recognize it also needs other essential elements as macronutrients and micronutrients. Above all, water, usually in liquid form, is a universal need and water requires physical and chemical conditions that allow it to an extent to retain its functionality as a liquid rather than as a solid (ice) or gas (steam).

Various forms of bacterial biomass have evolved to function high in the atmosphere, deep down in the hydrosphere, and within the Earth's crust. As surface dwellers, humans have become preoccupied with the surface biosphere that extends over the 30% of the Earth's surface area (designated terrestrial). This biosphere contains a vast variety of plants and animals that generally inhabit specific environments. A unique feature of these plants and animals is their ability to remain in one place and grow to obvious size and form (most plants) or move around to feed on other living organisms (animals). The oceans form another major surface-related environment for plants and animals. The plants tend to aggregate near water surfaces where solar energy permeates down from the sun in a diurnal manner (photic zone). Animals tend to find environmental niches throughout the ocean if they have significant levels of oxygen for respiratory functions.

The position of this atlas regarding the differentiation of prokaryotic bacteria is to concentrate on the most primitive of living organisms on the planet. All other organisms capable of self-reproduction are defined as eukaryotic and their cells contain many more structures than bacteria. For this reason, bacteria are considered one of the first groups of microorganisms subjected to the forces of evolution on earth and they are thus designated alpha group in recognition of the fact that they underwent the earliest durable development of any living organisms. Logical argument for the differentiation of the alpha group as dominated by bacteria is defined by the factors cited in Section 3.2.

3.2 DEFINITIONS OF ALPHA-BASED BACTERIAL CONSORTIA

Recent scientific endeavors naturally focus on the surface biosphere (200 m above and below the land and ocean surfaces). This creates challenges, particularly in relation to pathogenic risk (primarily for humans and secondarily to economically important exploited species) and cures (primarily for recognized pathogenic challenges and obvious or perceived environmental failures). The categorizations presented below are based on significant microbiological activities arising from the premise that liquid water is the major factor in initiating microbiological events. Bacteriologically influenced events along the viscidity gradient of liquid water on Earth—from the clouds in the atmosphere down to the super pressurized steam below—will be discussed.

Prime factors defining forms of significant microbiological activities were explained in Chapter 2: the oxidation–reduction potential (ORP) and viscosity or physical state of water (PSW) within the environment. Considerable evidence supports the manner in which ORP influences bacterial activities but much less recognition surrounds the primary influences of bacterial consorm activities on the characteristics of water. Water is found in liquid and bound forms based on temperature. Water can be measured by its viscosity and originates in four forms from a bacteriological perspective:

1. A solid, primarily in the form of ice comprised of complex crystalline lattices
2. A free liquid not encumbered with a significant burden of dissolved and/or suspended biocolloidal matrices

3. A bound form containing sufficient biocolloidal elements to affect physical characteristics such as flow and viscosity
4. Gaseous dispersed forms in which the water is commonly in a gas phase; may be found in the atmosphere

In the bacteriological communities, the state of water is possibly the most significant factor affecting the types of community structures that may develop. In this atlas addressing bacterial communities and their identification, the ORP and viscosity of water determines the types of consorms that may be active.

Common knowledge dictates that most bacteriological activity occurs in the free liquid water state. In reality, this may not always be the case. Bacteria often function most effectively within attached biofilms that grow over surfaces such as porous media (e.g., sands, clays, rock fractures), colloidal (gel-like) structures in bound liquid water matrices, or even aerosols containing water droplets of bionucleated particles. Bound liquid water may also exist within ice-generating sites that may involve bacteriological activities. The ice temperature would clearly be in the freezing range but bound liquid water trapped within the ice may form environments within which these bacteria may flourish. Because bound water creates conditions that are not restricted to the normal pressure-dependent freezing and boiling points of water, the potential exists for bacteriological activity well outside the normal range of 0 to 100°C water temperatures at normal atmospheric pressures.

Naturally, higher pressures could causing destruction of complex cellular structures that would then quickly die. Plants and animals both contain complex intracellular structures that make them very vulnerable to such pressure changes. Bacteria, by and large, lack complex intracellular structures and also have relatively flexible cell walls. These two features supply considerable forgiveness for pressure changes because bacterial cells compensate for the effects of changing pressures (such as hydrostatic head) and thus do not collapse and die.

The likely premise is that the extent of bacteriological activity within the crustal elements of Earth would not be restrained by increasing hydrostatic pressures generated by pooled surface waters or groundwaters in the crust or the oceans. Bacteriological activities have been detected in exploratory bore holes going as deep as 5 km and in abyssal waters more than 10 km deep. It is thus reasonable to surmise that the recovered bacteria functioned at very significant depths and pressures in a liquid (free or bound) water environment. This concept extends to bacteriological communities near the magma layer with the possible limitation at the interface where pressures and temperatures and are so high that water exists only as a gas and does not contain nucleations of liquid water in a form that could support biological activity.

It now becomes reasonable to assume that bacteriological activity may exist deep in the Earth's crust, and is defined here to exist down to the point of a permanent lack of bionucleating particles in the bound liquid water. This would mean that even the bionucleated gaseous steam there may support bacterial activity. From this steam-dense magma–crustal interface extending upward toward the surface biosphere the gridded atlas concept proposes six distinct gradients generated by thermal activity, salt concentrations, electrical activity, water content, shift of reductive to oxidative state when moving outward from the magma core to the atmosphere, with a reversed gradient reflecting movements of organic materials from the surface biosphere to the crust. These gradients all have different characteristics associated with concentration shifts and changes in electrical charge.

Electrical charges are created at least in part by the spinning action of the Earth that creates a dynamo effect. This concept is supported by several observations. First, the Earth has a strong magnetic field. Second, a strong gradient for electrolysis of water occurs throughout the water-saturated crust to create strongly oxidative regimes driven by the oxygen generated through bioelectrolytic functions at the same time hydrogen is generated to dominate the reductive conditions in the lower portions of the water-saturated parts of the crust. The third observation is a major diffusive downward gradient created by organics synthesized by the surface biomass driven by the availability of oxygen, presence of water, and adequate solar radiation.

While the primary function of organics is metabolic and reproductive within the viable bacterial cells of the biomass, the natural final consequence of growth is the release of surplus organics when the biomass dies. Releases of the bound forms of organics within the dead cell masses are then subjected to gravity-controlled movements that carry them down the gradient into the reductive zones of the environment. At the same time, some of these organics are subjected to oxidative degradation through releases of carbon dioxide and partial reassimilation of daughter products into cells in the biomass.

It is generally perceived that plants are the primary global generators of organics through photosynthesis while animals are the primary consumers of the plants. Thus, herbivores constitute the first level utilizers of organics from plants. Animals, however, are not efficient utilizers of these organics since such organics move up the food chain inefficiently as larger animals (carnivores) consume smaller animals. Inefficiencies are also created by the releases, primarily of semisolid organics through defecation, that over time far exceed the biomass weights of the living animals in question. These releases (as excreta) now become prime feedstocks for bacteria inhabiting the surface biosphere.

Organic feedstock for the sustenance of bacterial biomass comes from a variety of sources including plants (via direct decomposition of plant biomass), animals (via excreted organics and the decomposition of dead animal biomass), and microorganisms, particularly when the crust environment changes from oxidative to reductive based on depth. The primary concept is the differentiation of six major alpha bacterial consormial groupings based upon the reduction–oxidation state of the environment and the viscosity of the water in the environment.

Water is another key component that moves from generally dispersed (bionucleated or diffused) forms in the atmosphere to pooled collections on the surface of the crust where it can enter water-saturated porous media and fractures found at the interface of crust with magma. In the proposed classification of alpha bacterial consorms, the six groupings are defined as lateral biomass strata within the sections of the crust, surface biosphere, and overlaying atmosphere where liquid forms of water remain present. The conceptual differentiation of the six proposed bacterial consorms is based on the manner in which the environment becomes dominated by ORP (creating reductive or oxidative conditions; x axis); and the form in which water is bound (viscosity; y axis). Differentiation of the consorms into groups is therefore defined by a consorm's ability to:

1. Generate dispersed biologically nucleated water droplets (bionucleation leading to viscosity increases)
2. Form organic or inorganic coatings over viable cells (bioconcretions)
3. Interact with organic carbon and daughter products in reductive or oxidative states

The six proposed alpha communities are defined individually in Sections 3.3 through 3.8. Each functions as a distinctive part of an integrated overall cycle through which forms of liquid water support community structures. Each group is defined by its focal point location (FPL) and numbered (XX-YY, with x and y locators defined in the grid atlas). For example, fmv:fcP refers to the factorial positioning of the ORP (x axis) and the factorial positioning of the y axis in centipoise (cP). Thus, 22-04 represents grid position 22 (oxidative) for the ORP and 04 indicates water viscosity. The 22-04 combination means location in an oxidative region with water in droplet (dispersed) form.

3.3   ALPHA ONE: BIONUCLEATING DISPERSED CONSORMS [FPL (FMV:FCP) 22-04]

The alpha one community has not been widely explored or recognized and therefore remains wrapped in conjecture with very few hard bacteriological facts available. One common feature of alpha bacteria is their ability to nucleate water within charged particles suspended in a gaseous phase (commonly air), primarily as bound water droplets. Of all of the bacterial consorms, alpha ones are the most recognizable to a casual observer simply as the clouds in the skies. Clouds are seen as significant meteorological structures; they are not seen as bacteriologically driven entities.

What is so unique about this floating biomass driven by the winds is that it is generally very dynamic and in constant motion. Perhaps the inability of alpha one bacterial consorms to be cultivated on agar surfaces led to the premise that rains and clouds are sterile. In reality, clouds, rain, and even snow are not sterile; monitoring protocols simply failed to detect the bacteria they contain. Testing, particularly with the HAB-BART™ (Heterotrophic Aerobic Bacteria-Biological Activity Reaction Test) system in combination with a visual BART reader (VBR72) allows detection of very active bacterial populations within these natural materials.

Completed cycles generated by alpha one bacterial consorms commonly end in precipitation as rain, snow, or droplet dispersion forms such as mists and fogs. These events may also involve electrical discharges associable with inter-reactive latent electrical charges that may build up within neighboring clouds and surrounding atmosphere. So dynamic are these clouds that complete cycles can occur in minutes, hours, or days. Close connections of the various alpha one consorms and the surface biosphere (dominated by alpha two and alpha three forms) exists because the microorganisms suspended in the bionucleated clouds fall to the surface with precipitation and hence are "inoculated" into surface waters and soils.

Conversely, wind and convection currents in the atmosphere move bacteria from the surface waters and soils within dispersed convected suspensions and aerosols to again trigger atmospheric bionucleating events. It may be expected that, while suspended essentially in the clouds, these bacteria will continue to perform primary functions including the selective accumulation, storage, and breakdown of volatiles, suspended organics, and other molecules moving upward into the atmosphere from the surface. Such a sequence of events means that clouds may trigger events parallel to bacteriologically influenced events in soils and water associated with bioaccumulation and degradation functions.

Alpha one bacterial consorms essentially act as floating natural filtration systems in the atmosphere through which chemicals pass as they move from surface environments toward deep space (if not accumulated and degraded). A significant example of this type of bacteriological activity is the reported damage to the ozone layer considered to have been created in part by man-made organo-halogen compounds such as chlorofluorocarbons (CFCs) and bromofluorocarbons (BFCs). Examination of the diminution of the ozone layer around the North and South Poles revealed considerably more impact damage above the polar regions of the Southern Hemisphere. It is interesting that roughly two thirds of the land surfaces on Earth are in the Northern Hemisphere. This means that the stratosphere in the Northern Hemisphere faces much higher exposure to air passing over land masses that may impart more nutrients into the atmosphere where they could become incorporated into alpha one bacteriologically influenced clouds. Such additional incorporations could then act to stimulate incumbent bacterial communities in the clouds to greater accumulative and degradative activities. For CFCs, greater rates of degradation could occur in the more eutrophic clouds of the Northern Hemisphere, resulting in fewer secondary impacts on the ozone layer positioned above the atmosphere. At the same time, the Southern Hemisphere would generate more oligotrophic conditions that support fewer dynamic bacterial activities; more CFCs would escape to damage the ozone layer.

We must determine the natures of clouds from a bacteriological perspective since these alpha one communities may act as Nature's floating fluffy Band-Aids and perform critical functions related to the meteorologies of surface environments. Comprehensive understanding of the alpha one consorms would have particular value for determining the significant factors affecting the dynamics of global temperature changes at all levels.

3.4 ALPHA TWO: ORGANIC BIOCONCRETING CONSORMS [FPL (FMV:FCP) 22-16]

These bacterial consorms may be commonly defined by their ability to form predominantly organic matrices outside cells that are commonly polymeric in nature and have the ability to bind water within these extracellular structures that may be attached to solid or liquid surfaces. Scientifically these growths are called biofilms since they are films attached to surfaces and generated from glycocalyces; they are more popularly known as slimes. These slime growths are commonly found in the surface biosphere. Unlike plants and animals, whose major structures are synthesized within cells, bacteria tend to generate few structures inside cells but are created mostly outside cells. These structures generated outside of the cells tend to have concretious forms based on the binding of polymeric organic molecules to water (hence the greater viscosity). The net effect of these structures is to provide a type of loose "skin" of bonded water that fills protective, accumulative, and storage functions for incumbent cells.

Thus, unlike plants and animals, alpha two consorms increase their biomasses primarily outside cells rather than inside. These bacteria play major roles in surface environments such as soil and water and can also generate parasitic functions that may or may not be deleterious to an infested host. Parasitism may be nondisruptive to a host and may also be mutually advantageous to parasite and host. However, if a parasitic function becomes pathogenic to a host by causing serious damage or lethality, the relationship is disrupted. It may be reasonable to surmise that bacterial pathogens (generally single strains) are as disruptive to a parasitic bacterial consorm as to a host. In the development of a harmonic state between host and parasites, a pathogen not only impacts the host but also the bulk of the bacterial biomass active within the host!

The FPL of alpha two consorms is (fmv:fcP) 22-16. This defines the consorms collectively as functioning most efficiently in an oxidative regime (fmv 22). and requiring more viscid water (fcP 16). Most of these consorms therefore tend to bind water using the polymeric network around cell exteriors. Hence these consorms usually favor oxidative conditions within biofilms or slimes. It is not in the nature for the alpha two consorms to accumulate metals and most growth takes the form of attached biomass.

3.5   ALPHA THREE: INORGANIC BIOCONCRETING CONSORMS [FPL (FMV:FCP) 13-21)

These bacterial consorms commonly occur within the water-saturated crust of the Earth, particularly in the zones between the oxygen in the saturated water and the zones where oxygen is absent—at the oxidation–reduction interface where major bacteriological activity is focused. At the interface, chemical utilization by bacteria shifts from a reductive form (in the water-saturated crust) to the oxidative environmental usually associated with oxygen from the atmosphere. A number of bacteriologically influenced concretious events occur at the oxidation–reduction interface. For example, iron moves from relatively soluble reduced (ferrous) forms to insoluble oxidized (ferric) forms, often in the form of ochres. Uncontrolled bioaccumulation events produced by alpha three consorms may lead to bioconcretions containing high concentrations of oxidized cations. Examples include the formation of ferric iron pans at the oxidation–reduction interface and the generation of iron-rich rusticles of the types observed on sunken steel ship wrecks.

Increasing evidence indicates that even elemental gold may in fact be bioaccumulated by bacteria at oxidation–reduction interfaces. Additionally, dolomite limestone may be the result of complex bacteriologically driven electrolytic activities including those of alpha three consorms. These types of inorganic bioconcretions may also be major factors in the formation of sedimentary rocks and synthetic concretes. The polymeric crystallised matrices generated by the community primarily accumulate as oxidized inorganic elements leading potentially to greater structural integrity.

3.6 ALPHA FOUR: CARBON-REDUCING CONSORMS [FPL (FMV:FCP) 06-27]

Organic materials created via the functional dynamics of a biomass form a part of a cycle that may involve three elements:

1. Plants as primary generators of organic carbon
2. Animals and microorganisms as prime degraders of organic carbon under oxidative conditions
3. Bacteria-reducing organics under reductive conditions to molecules dominated not by fatty acids ($C_XH_YO_Z$), but by hydrocarbon molecules (containing only carbon and hydrogen)

While the degradation of organics in the surface environment may generate carbon dioxide as a primary end product, when organics enter saturated reductive environments in the crust the alpha four bacterial consorms dominate. Essentially this triggers a biochemical stripping process in which potentially important nutrient elements (e.g., nitrogen, phosphorus, sulfur) and oxygen are removed, leaving molecules that contain only carbon and hydrogen. Four terminal carbon depositions may be associated with this reductive stripping.

First, the organics are reduced to fatty acids (C_XH_YCOOH) under the R (very reductive) and G (moderately reductive) conditions shown in the atlas grid map if sufficient water is available. This action appears to be the primary reductive intermediate driver for organics formed in the surface biosphere that migrate downward into the reductive parts of the Earth's crust.

The second stage occurs in the downward migration. The organic molecules dominated by fatty acids are further bacteriologically reduced to long-chain hydrocarbon polymers commonly with 8 to 50 or more carbon atoms. These form the crude oil deposits commonly observed in deeper geological strata subjected to organic intrusions.

The third state involves a reduction of organic carbon involving further bacteriological reductive activities that reduce the hydrocarbon chain lengths to 1 to 3 carbon atoms with the single-atom methane (CH_4) commonly dominating. Certain specialized bacteria can generate methane biogenically as a terminal gas product; they are known as methane-producing bacteria. Generally biogenic methane is considered to be generated only at shallower depths and lower pressures while methane with a higher isotopic ^{13}C content is termed thermogenic because the conditions under which it is produced hare considered too extreme for bacteriological activity. This concept remains to be critically evaluated.

The fourth reductive stage exists when organic carbon is reduced to elemental forms, primarily as various grades of coal. These coal deposits are often relatively shallow and may reflect bacterial activities at redox fronts established in association with the upper zones of saturation in the porous crustal media (G or moderately reductive condition based on the atlas grid). Significant electrolytic effects may occur; primitive laboratory trials involving electrolytic functions in unconsolidated porous media frequently generate hydrocarbon or gaseous materials around the anodes. In these reduced forms, these daughter carbon-rich molecules remain stable until moved back into the oxidation–reduction interface where oxidative biodegradation may occur.

3.7   ALPHA FIVE: CARBON-OXIDIZING CONSORMS [FPL (FMV:FCP) 13-07]

In the deeper reductive zones saturated with water inside porous and/or fractured geological media, alpha four bacterial consorms reduce the organic molecules to various combinations of carbon and hydrogen. Dominating products range from simple gases such as methane to complex mixtures of petroleum hydrocarbons. Coal as elemental carbon represents a more reduced form. In the reduced state, these compounds remain relatively stable but become subject to bacteriological oxidative degradation when the molecules move upward into conditions associated with the redox front (grid condition F) and the atmosphere (grid condition O, fundamentally oxidative).

Alpha five consorms are dominated by carbon oxidizing functions. They can attack hydrocarbons by oxidatively degrading the longer chained petroleum hydrocarbonaceous molecules. Methane is typically subjected to rapid degradation through oxidation. Alpha five consorms are typically active at the oxidation–reduction interfaces in soils, waters, and sediments and are capable of degrading the smaller hydrocarbon molecules at the sites. Gas hydrates positioned just below the ORP interface (grid condition F) are commonly found in marine sediments and under permafrost illustrate the role of the alpha five communities in the degradation of bioaccumulated methane trapped within hydrates. The lower part of a gas hydrate performs the role of a reductive bioaccumulator, entrapping methane efficiently within an ice lattice. In the upper parts of the gas hydrate, at or above the more oxidative (grid condition O) zone, oxidation of organic carbons and production of carbon dioxide occur. The FPL of the alpha five consorms (fmv:fcP 13-07) reflects the fmv gradient where the environment is just becoming oxidative and the water is not encumbered chemically and biologically.

3.8   ALPHA SIX: HYPERBARIC DISPERSED BIONUCLEATING CONSORMS [FPL (FMV:FCP) 01-03]

The deep interface between the hot magma and groundwater moving downward through fractures and porous geologic media of the crust of the Earth remains virtually unexplored. In this zone, a transition between liquid water under extreme pressure and gaseous steam occurs. It is reasonable to conjecture that transitory environments may exist in which nuclei of bound liquid water aerosols within which bacterial activities may occur in a manner determined during the evolution of the planet. If this were supported by scientific investigation, a parallel consortial structure to alpha one in which bacteria adapted to these very extreme environments of high pressure and temperature could be proposed. At this time the proof of existence of such consortia remains conjectural but confirmation may be obtained via searches for alpha six bacterial consorms in seeps, vents, and groundwater flows. Proof may have to be achieved by a demonstration that these alpha six consormial bacteria may be cultivated in these hyperbaric environments at high temperatures and only when liquid water may still be present.

3.9 SUMMARY

This lumping technique for the differentiation of bacterial consorms proposes a division into six groups on the bases of different uses of forms of available water. Many of these bacteria exhibit a natural potential to generate polymeric structures in the environmental matrices surrounding living cells. Each of the six alpha bacterial community groupings appears rational but proof in some cases remains speculative. For alpha six consorms, the major scientific challenges remain associated with the recovery of suitable samples and development of cultural and biochemical confirmatory determinations. Table 3.1 summarizes the relationships and locations of all six divisions.

Using this lumping approach to the classification of bacteriological consorms provides distinct advantages over the traditional Linnaean approach of speciation primarily at generic and species levels based on replicable commonality of characteristics. In the natural world, it is common to observe these groups (communities of species) functioning together to achieve the common goals of creating biomass,

TABLE 3.1
Relationships of Proposed Alpha Bacterial Consorms

Community	Atmosphere	Soil-Based Biosphere	Water-Based Biosphere	Water-Saturated Crust	Magma–Water Interface	FPL (fmv:fcP)
Alpha one	+++a	–	–	–	–	22-04
Alpha two	+	+++b	+++c	++	–	22-16
Alpha three	–	+	–	+++c	–	13-21
Alpha four	–	–	–	+++d	–	06-27
Alpha five	+	+++e	+++e	+	–	13-07
Alpha six	–	–	+	+	+++f	01-03

Abbreviations: +++ = dominant and diverse. ++ = present and dispersed. + = present. – = absent. a = little confirmation of size and diversity. b = broad range of pathogens; most extensively studied of alpha communities. c = dominant at reduction–oxidation interfaces (redox fronts) in water-saturated porous and fractured geological media, particularly where metallic cations accumulate. d = limited to very reductive water-saturated media with organic permeation; still subject to confirmatory studies. e = occurring at a multiplicity of primarily oxidative environments as the reduced carbon (as gases and volatile)s migrates up to the surface. f = speculative postulation not yet supported by hard scientific proof. FPL = focal position locator applicable to 24 × 30 grid map of atlas.

reproducing, and developing a protective and defendable environmental niche from which colonization can be initiated. Unlike plants and animals, the alpha groups of bacteria rely upon communal intelligence that allows differentiation of roles among various members of the consorm for purposes of survival, growth, and dominance within the colonized ecosphere. This approach runs directly against classical Linnaean concepts of classification but may offer an easier method to determine the various roles of bacteria in the natural world. Evidence now indicates that bacteria dominate the global biomass and we must understand the dynamics and intelligence of these communities.

4 Preliminary Differentiation of Alpha Bacterial Consorms

4.1 INTRODUCTION

This approach to the identification of bacteria in natural environments involves two important basic premises: (1) bacteria only function effectively as integrated community structures defined as consorms; and (2) no single strain dominates such consorms although traditional methods of identification may suggest that.

Exceptions to these conditions may occur under abnormal conditions that may be experienced when a single strain dominates a community to the detriment of the impacted host (animal, plant, or microbial). Such a condition would create a distortion of the natural functions of the animal tissues, plant tissues, or microbial community that materializes as a dysfunction (disease). Strains causing these types of effects may be considered pathogens and are very disruptive to normal tissue cells and microbiological processes. Pathogenic bacteria, by their very nature, may be considered counterproductive in a potentially terminal manner to the normal functioning of a consorm.

This approach essentially "lumps" the bacteria into groups based upon common focal centers for specified activities. The alpha system of identification places emphasis on the categorization of normal bacteriological communities observed as active within consorms rather than on the dysfunction created by pathogenic activities. This chapter defines the major consorms; Chapter seven defines the consormial functions, habitats, and forms in more detail.

It is proposed that the bacterial kingdom, as defined by consormial activities, be split into the six major alpha groups defined in Chapter 2. This chapter develops a mechanism to differentiate the major consorms by their activities and reactions rather than from extrapolations based on isolated bacterial strains recovered from consormial samples. The prime mandate of this approach is to consider individual bacterial strains to be of less significance than communities of strains as a whole that jointly function in an reactive, intelligent, and integrated manner. Past practices of isolating monocultures are now counterproductive because the act of selection also means rejection of unculturable (uncultivatable) bacterial strains that are essential to a consorm. These strains that were not recognized by isolation have been ignored in the development of the classification hierarchy. Each of the six alpha types will be addressed and a more specific atlas location assigned to the recognizable consorms.

To differentiate the consormial groupings, the sequence explained in this chapter is organized primarily by alpha group (1 through 6), secondarily by the grid focal point location (FPL) on the atlas, and finally by a three-letter coding applicable specifically to each consorm (e.g., BPL indicates black plug layer). It is important to recognize that each bacterial consorm described is subject to methods of self-governance and self-regulation with highly flexible reactivity to the two-way interaction between consorm and embracing environment. Unlike static cultured pure bacterial cultures in a laboratory, the characterization of a consorm can vary over time, leading to shifts in the dominant strains and the manner in which the consorm functions. The FPL set by first by alpha group number (1 to 6) followed by fmv (oxidation–reduction potential or ORP), a hyphen, then fcP (water viscosity).

4.2   ALPHA ONE: BIONUCLEATING DISPERSED CONSORMS (FPL 1, 22-04)

Alpha one bacterial consorms have a common ability to nucleate water within charged particles suspended within gaseous, liquid, or solid matrices. This can take a number of structural forms but the primary premise is that the nucleated water in some way involves extracellular polymeric substances as the prime drivers of the bonding mechanisms in the isolated liquid water phase. In all cases this bound liquid water now forms a primary structure that can be dispersed in air; forms boundaries for trapping gases in close association with liquid water interfaces; or maintains a liquid-bound water environment within a solid water (ice) structure. The seven primary alpha one bacterial consorms are separated based on the definitions above and will be defined in Section 4.2.1.

4.2.1   Principal Alpha One Consormial Activities

1, 17-15 (CMT)* — Comets are, for the purpose of this definition, limited to ice comets that move through deep space. They are predominantly composed of ice within which it is conjectured that various microorganisms may survive for prolonged periods in the hostile environments of deep space.

1, 22-03 (CLD) — "Cloud" is the common name applied to all masses of dispersed water that float at different altitudes in the atmosphere. While it is well understood that clouds are the sources of rain and snow, their true microbiological nature remains to be fully explored. Sometimes called Nature's floating Band-Aids, they are in fact complex microbiologically influenced structures that bind (nucleate) water. Binding is a result of a combination of chemical polymeric and electrical charge activities that gives clouds the ability to float and maintain structural integrity at certain densities. Chemicals moving up into clouds are likely to be subjected to microbiologically influenced accumulation and degradation functions.

1, 19-06 (FO) — Foam is a collection of bubbles collectively formed in an integrated manner at the water–air interface and maybe near solid surfaces floating on water. Foam is essentially a coalescence of hemispheres (bubbles) composed of biofilms (slimes) within a water film holding entrapped gases inside the bubbles and away from the atmosphere above. These gases may be recalcitrant (e.g., nitrogen) or

---

* Speculative.

degradable (e.g., methane). Degradable gas bubbles would be expected to collapse quickly as the gas is degraded.

Bubbles are products of the entrapment of gases in water to form hemispherical (dome-like) objects that rise to the water surface where they may remain for some time. The gas in the bubble can include air (if the air is entrapped in the water through turbulence such as wave action), carbon dioxide (from respiratory activities), methane (from reductive degradation of organics), nitrogen (from the reduction of nitrites and nitrate), and hydrogen (through metabolic and electrolytic activities). Bubbles most commonly form in biocolloidal waters where considerable bacteriological activity leads to extracellular polymeric substance (EPS) production that serves to bind the water. The rapid formation of a skin around the gas bubble is usually dominated by biofilms. When a gas bubble rises to the surface of the water, the biofilms will remain intact and create foam. The gases in the bubbles may be used by bacteria within the biomass.

1, 16-12 (ICE) — Ice is solid water that usually takes a complex crystallized form (e.g., snowflakes) or forms a solid. Bacteria can survive and metabolize within ice by creating polymeric "antifreeze" that keeps some of the ice in a liquid state as bound liquid water in and around the bacterial biomass. This allows the bacteria to remain active within the ice at temperatures as low as −18°C.

1, 16-03 (LNG)* — Lightning is a releases of powerful electrical discharges observed primarily in the atmosphere. The three basic forms are (1) regular negative; (2) more powerful positive; and (3) sprites that tend to form above cloud tops. Bacteria associated with clouds are negatively charged and are capable of causing the buildup of charges within the infested clouds that can lead to discharges as lightning. While it is well known that bacteria can be intimately involved in the bionucleation of water in clouds, their role in the generation of microbiological capacitors, transformers, and transistors within clouds remains to be established.

1, 19-12 (SNO) — Snow is composed of crystallized forms of ice generated in clouds when subjected to freezing temperatures. One origin for the formation of these ice crystals forming followed by their precipitation as snow is the role of EPSs formed outside bacterial cells in binding water as crystalline solid ice when temperatures drop. Essentially the freezing of water occurs relatively quickly, leading to increased density of crystallized snowflakes. The form that a snowflake takes during ice nucleation is a response to the polymeric structures present in the bacterial consorm. Commercial snow making commonly employs specific polymers from known strains of bacteria (particularly Pseudomonas genera). The snowflakes generate common crystal patterns of ice via freeze nucleation of water droplets in clouds.

4.3　ALPHA TWO: ORGANIC BIONUCLEATING CONSORMS (FPL 2, 22-16)

Most bacteria we encounter daily are associated with the surface biosphere at the oxidation–reduction front where organics generated primarily by plants move

---

* Speculative.

downward into water-saturated zones. Essentially, much of the organic carbon is synthesized primarily from carbon dioxide by direct photosynthetic mechanisms of the carbon cycle. Once in the organic form, carbon acts as a driver for the synthetic and catabolic process associated with life. Alpha two bacteria are essentially involved in processes associated with the maintenance of the organic fractions of the biomass. Organic carbon constitutes a dominant part, frequently gravimetrically reaching 50% of total dried chemical content. Alpha two bacteria are therefore predestined to accumulate and recycle various forms of organic carbon and thus contribute to life on Earth. The bioconcretions are dominated by organic forms of carbon.

4.3.1   Principal Alpha Two Consormial Activities

Most alpha two bacterial consorms have three common abilities: (1) they can accumulate bound water; (2) they are active within a relatively narrow ORP range in the selected R, G, F, or O segments of the gridded atlas; and (3) they often compete with other life forms (plants, animals, and microorganisms) while generally exploiting a relatively narrow environmental niche. Alpha two bacterial consorms dominate only in conducive environments and lack flexibility to adapt to other environments. However, these consorms are robust and resilient and can be flexible and adapt to fresh environments by modifications of the roles of incumbent bacteria. Essentially alpha two bacteria do not have mechanisms for prescribed death or senescence.

2, 12-17 (ABS) — Abscesses are holes in animal tissues resulting from microbiologically influenced activities that include the destruction of tissue cells and structures, generation of pockets of gases, and pooling of bacteriologically rich fluids. They are recognized by swelling of tissues (due primarily to gas formation) and high body temperatures (due to bacterial activity). Abscesses may rupture internally or externally and pus (surplus gas plus bacteriologically rich colloidal waters) may be released.

2, 13-10 (BOL) — Boils are bacteriologically influenced infections in or near skin tissues. They produce skin swelling from pressure at the infection site. Tissues at the site become badly infected, to the extent that the local blood and lymphatic systems are plugged. This prevents gas or fluid pressures from being relieved. As a result, a dome-like growth rises through the surfaces of the skin. A boil may be relieved when the skin breaks from the pressure and bacteriologically rich fluids flood from the infestation site.

2, 19-07 (BCH) — Bronchitis refers to common forms of bacteriologically influenced lung infections. Air passageways become impacted by both biofilms forming on tissue surfaces and bacteriologically bound water, both of which restrict the movement of oxygen in the air into and carbon dioxide from the lungs. Severe infections can weaken the immune response centers and lead to respiratory failure, usually from secondary infection causing pneumonia.

2, 08-20 (CBC) — Carbuncles are deep-seated infestations within warm-blooded animal bodies. Tissue swelling may also involve complex bone structures and muscles. The net effect is a swelling of the tissues under intense pressures generated by the bacteriologically influenced blockages of the venous and lymphatic systems. Because of the depth of carbuncles, treatments become more challenging.

2, 11-14 (CCD) — Carietic conditions arise from bacteriologically influenced corrosive effects primarily on tooth enamel but secondarily on the dentine and gums, eventually causing infected teeth to rot. Associated with the secondary impacts are tissue swelling, plugging of the venous system, and collapse of the nervous system along with chronic or acute pain.

2, 15-07 (CLS) — Cloudiness is weather that lacks clarity. When cloudy water is held up to a light, microscopic colloidal particles will be seen reflecting and/or absorbing the light to produce a cloudy appearance. Such cloudiness is commonly biocolloidal in form and may at times appear as a series of lateral layers floating in otherwise clear water. In water wells, the lateral layers may concentrate at the oxidation–reduction interfaces that form between oxidative and reductive conditions.

2, 11-18 (CHR) — Cholera is a disease that exerts a major impact, particularly in the gastroenteric tracts of humans. Symptoms start with profuse watery diarrhea, followed by vomiting and leg cramps. If the infestation continues, a patient may suffer a rapid loss of body fluids, leading to dehydration and shock.

2, 16-13 (CLB) — Coliform bacteria appear in the gastroenteric tracts of animals. They have a natural resistance to bile salts and utilize lactose as a principal energy source. They are important indicators of the contamination of potable waters with fecal material. In the 20th century, coliform bacteria were used successfully to determine hygiene risks associated with diseases such as cholera, dysentery, and typhoid (three major causes of decimation of humans, particularly during famine or drought).

2, 15-18 (CLW) — Colloidal water is free liquid water that generally contains colloidal elements created by bacteria by generating polymeric matrices from EPS which then bind some of the water. Water in biocolloidal form viscosities different from those of free liquid water and generally contain significant suspended solids with small diameters in the range of 1 to 35 microns. The bacterial communities tend to be clustered in the water within the biocolloidal elements and appear as floating structures with relatively distinct edges.

2, 18-13 (DRR) — Diarrhea is the passing of increased amounts of loose stool material that can lead to dehydration. It is often caused by bacteria and can be acute (short term) or chronic (long term; >2 to 3 weeks) in humans. More children die of diarrhea arising from of poor sanitation than any other disease.

2, 10-12 (FCB) — Fecal coliform bacteria can survive bile salts, ferment lactose, may be incubated at 44.5°C, commonly inhabit human and animal intestines; some can cause clinically significant infections. Originally it was thought that cultural conditions would limit detection to *Escherichia coli* but a number of other species within the Klebsiella and Enterobacter genera are also detectable on occasion. In the past two decades, more attention has been paid to the specific detection of *E. coli*, generally considered the best marker for hygiene risk from water and food.

2, 07-23 (FEC) — Feces are solids excreted from the gastroenteric tract and include a very diverse range of bacteriological communities that function in different parts of the tract. These communities have a common purpose: aiding the degradation of foods and liquids consumed by the host. Essentially such communities are the "first feeders" of ingested food and the daughter products of their activities are then absorbed by the host through the walls of the gastroenteric tract. Fecal material excreted from the body tends to contain relatively little water and is bulked by the living biomass and recalcitrant materials (e.g., fibers) that cannot be digested. Because the intestinal tract acts as a direct conduit through the body, many pathogens inhabiting the tract are excreted with feces. This is the prime reason that hygiene risks are linked sites where feces are deposited or dispersed.

2, 10-08 (KDS) — Kidney stones are inorganic solids that occur primarily in the kidneys and urinary tract. They may be generated as daughter products of carbonaceous bioconcretions generated by bacteriological activity at those sites.

2, 22-12 (LMN) — Luminescence is a condition by which light (usually in the blue and ultraviolet range) causes colors to be generated (such as on water surfaces and encrustations) or through the generation of light within a biomass (bioluminescence), usually as a result of the release of photons by the discharge of high energy phosphorus (adenosine triphosphate, ATP) triggered by a luciferase catalyst.

2, 09-15 (MIC) — Microbiologically influenced corrosion refers to losses in structural strength and integrity, primarily of steels although other metals and materials such as concrete may also be affected. MIC takes on a number of forms but fundamentally microbial activities remove metals and other materials at a metabolically impacted site, causing lateral spalling, pitting, and eventual perforation. Two bacterial groups commonly associated with MIC are sulfate-reducing bacteria (SRB) that generate hydrogen sulfide (causing electrolytic corrosion) and acid-producing bacteria (APB) that cause acidulolytic corrosion. The MIC consorm falls in the alpha two group because the initiation of a growing biofilm leads to corrosion and triggers the secondary events leading to the corrosion of colonized surfaces. MIC therefore begins as an alpha two consorm and then adapts to the accumulations of materials common to alpha three consorms.

2, 18-16 (NCR) — Necrosis is a condition causing plant or animal tissues to die because of some infestation or physiological failure. Generally necrotic conditions can be recognized by impacted tissues that dry out (lose the ability to hold bound water) and darken in color.

2, 16-09 (ODR) — Odors arise from olfactory (smell) recognition of daughter products from microbiological activities. Some alpha two consorms generate odorous products. These odors can be grouped as: (1) septic, commonly associated with the reductive breakdown of organics and the releases of volatiles; (2) rotten egg odors associated with the generation of hydrogen sulfide; (3) sweet odors (e.g., from apples, bananas, fresh vegetables and fruit) are often associated with the oxidative activities of bacteria, yeast and molds; (4) rotting vegetation odors often associated with the reductive activities of bacteria within organic-rich environments.

2, 22-21 (PMP) — A pimple is a dome-like swelling of the skin generated in part by localized bacteriological infestation. Skin eruptions may arise from hydrostatic and gas pressures resulting from pimple growth. These eruptions relieve the pressures by generating and releasing viscid or watery pus.

2, 19-21 (PLG) or 3, 19-21 (with bioaccumulation) — Plugging is created by biomass that grows within fractures or water pathways (such as through sands and pipes) and causes reduced water flows and/or diversions. This biomass will often form at oxidation–reduction interfaces and may also contain accumulated ferric iron, synthesized carbonates, and entrapped materials (sands and clays). As a result, the biomass may become dense and rigid. Such forms are often referred to as ochres or iron pans and reflect historical sites where redox fronts have formed. Plugging usually starts as relatively soft material dominated by active biomass and treatment can be applied effectively. However, as the biomass matures and becomes denser and more rigid, treatments become more challenging.

2, 23-14 (PEY) — Pink eye is characterized by a pink to red eye color that may result from swelling of the blood vessels in the eye or may be associated with a bacteriologically influenced infestation that may include the generation of red pigments by the bacteria.

2, 16-18 (PUS) — Pus is a viscid fluid exudate from a bacterial infection that erupts under pressure. Such exudates occur in maturing boils and pimples under pressure. Not so well known are other settings in the natural world such as tubercles—dome-like growths fracture at the tops and release bacterial biomass inside the domes. Because of the high iron content of the domes these exudates are very viscid and often appear as extensions of a growth before dispersion of exudates can occur.

2, 06-18 (ROE) — Rotten eggs result from degradation of the contents of avian eggs. The shell is not involved; the major impact is on the albumen and/or the yolk. Eggs are protected by outer shells that form impermeable barriers. Lysozymes contained in egg contents retard the ability of bacteria to become active after gaining entry. Three types of rot are well recognized: black rot (including the generation of hydrogen sulfide), red rot (contents turn red as a result of bacteriological activity), and green rot (contents liquefy and often glow in ultraviolet light).

2, 14-15 (ROT) — Rot is a condition by which bacteria enter animal or plant tissue cells and cause a general collapse of tissue structures; it may involve the collapse of tissues into a mucoid amalgam of bacterially dominated activities. Of the various rots, carrot root rot by species of Erwinia is the best known.

2, 16-21 (RZZ) — Rhizosphere reflects major bacteriologically dominated environments surrounding plant roots in moist or saturated soils. This zone around any plant root (weed or tree) is the site of interactions of roots and soil—the site where plant roots "barter" with soil microflora for a two-way exchange of water, nutrients, and oxygen. Such interactions may be very integrated: microorganisms moving into plant root tissues and various growth stimulants released by plant roots. It should be noted that microorganisms other than bacteria can also play major roles in the rhizosphere surrounding plant roots.

2, 22-07 (SHN) — Sheen forms on the surfaces of water and other materials when ultraviolet fluorescent materials are present. The fluorescent materials may be bacteriological in origin. When sheen is observed on a surface (such as a still pond) most likely an active biofilm at that air–surface water interface contains the bacteria generating the pigments.

2, 23-21 (SKN) — Skin is a tough outer layer that protects the biomass inside. Human skin and slimes are examples of this tough protective coating. However, "skin" may apply to any condition by which the surface of a liquid becomes denser and more rigid. Such surfaces are commonly bounded by air on one side and some form of water on the other. A greater amount of focused oxidative activity occurs in such skins. When a biomass forms a tight skin, the treatment to disperse the biomass becomes much more challenging.

2, 11-16 (SLT) — Silt in geology, engineering, and agriculture refers to particles that are smaller than sands but larger than clays. It is not well known that silts are often rich in bacteriologically dominated biomass and reflect forms of suspended and dissolved matter where considerable retention of bound water and bioaccumulation occur within the silt consorm.

2, 21-22 (SLM) — Slime arises from glycocalyces that define the amount of bound water that forms around the outsides of bacterial cells. This water is held in place by polymeric weaves of extracellular polymeric substances (EPS) created primarily by bacteria as protective envelopes (or skins) around cells. Slime commonly attaches to young biomass and the growth and appears very slimy. Many colors may be generated based on pigments produced by the bacteria or accumulates (e.g., ferric iron may dominate yellow, red, or brown colors in the slime).

2, 15-13 (TBD) — Turbidity occurs when crystal clear water becomes exceedingly cloudy. The cloudiness arises from suspended particles that include some elements of dispersed biomass in the water. Cloudy (turbid) water may be unsafe to drink because its high level of suspended solid content may contain hygiene-risk bacteria.

2, 24-13 (TCT) — Thatch is a weave of dried plant material designed to provide protection from the elements. In another context, many bacteria infest plant leaves (e.g., turf grass) by "gluing" the leaves to each other to form a canopy using EPS generated by the bacteria. Thatch will divert rain to the edge of the infested site so that the soil under the thatch may become very dry, causing the thatch-infested plants to die back. This is commonly called "wind burn" but is actually a bacteriological infestation of leaves.

2, 07-20 (ULC) — Ulcers are found within the gastroenteric tracts of animals and may be caused by localized infestations of surrounding tissues. When bacteria become involved, localized tissue reactions may cause growths (polyps) or collapses (necrotic lesions) within the impacted tissues. Traditionally these conditions have been diagnosed as cancers (alpha two ulcer consorms are often challenging to culture). The scope and forms of ulcers are not yet fully understood.

2, 12-07 (URN) — Urine is theoretically a sterile liquid excrescence of waste liquids and dissolved solids from an animal. Although generally considered sterile, urine may become the site of infestation (urinary tract infection or UTI) in the bladder, renal tract, and kidneys by bacterial consorms.

2, 12-13 (WLT) — Wilt can be observed in vascular plants under conditions when the vascular system (composed primarily of phloem and xylem that function as conduits for nutrients and water) becomes plugged by a bacteriological infestation. All plant tissues above the impacted region will become water-starved, causing collapsing (wilting) conditions. The head of the plant droops first due to the loss of water to maintain internal pressures. This wilting phenomenon continues as the plant collapses above the site of the infestation and then dries out.

4.4 ALPHA THREE: INORGANIC BIONUCLEATING CONSORMS (FPL 3, 13-21)

Alpha three bacterial consorms differ from alpha two communities in that the biomass generated is not necessarily dominated by organic forms of carbon, but rather by accumulations of other commonly inorganic chemicals. Alpha three bacterial consorms therefore tend to follow organismal life cycles because the accumulated chemicals ultimately dominate the biomass to the detriment of the consorm. However, alpha three bacterial consorms are innovative and consistently migrate to new and more acceptable environments in which to grow.

Each alpha three bacterial consorm focuses on the gathering and accumulation of specific chemicals (such as iron or carbonates) and operates as a chemically closed system—the biomass does not excrete significant metabolic wastes. In general the living biomass as a percentage of the total mass declines with maturation while the dead accumulates increase within the gross biomass. After the critical point is crossed and the biomass is spent (can no longer function) it is subject to outmigration by the cells in search of new habitats. Bioconcretion in these consorms relates to the intelligent accumulation of inorganic materials that provide support and protection for the associated biomass.

4.4.1 Principal Alpha Three Consormial Activities

Alpha three bacteria have the capacity to accumulate particular inorganic compounds well beyond their metabolic needs. This causes an increase in the density of the biomass with a parallel decrease in the density of viable cells within the biomass. This means that alpha three bacteria do not follow a life cycle ending in the deaths of cells remaining within the biomass (senescence) like most organismal life forms. Migration of cells from a senescing biomass leads to the formation of fresh alpha three biomass growths at environmentally suitable sites. Alpha three consorms usually form at selective ORP sites (particularly in the G and F ranges on the grid atlas), interact, and control the viscosity of the water within a colonization site.

Accumulates normally found in alpha three consorms involve the transition of ferrous (soluble and reduced) iron to the ferric (insoluble and oxidized) state that dominates many alpha three consorms. This occurs over a factorial range of ORP from 13 to 18 fmv; generally the colors of consorms growing under these conditions vary through orange, red, brown, and even black, depending on the age of the biomass and the availability of iron. Other ORP values present the potential for the accumulation of other metallic cations.

Biomass forming along an ORP gradient may be able to differentially accumulate different cations and can produce a "slime" chromatograph in which the different cations are collected at different ORP values. Around a water well, as an example, a differential accumulation of metallic cations occurs along the biomass gradient from oxidative (near the well) to reductive (further away from it). A typical bioaccumulation sequence for captured cations (moving first from the oxidative end of the interface, then becoming more reductive) is Fe < Zn < Cu < Mn < Al < Ti < Au < As < Pb < Cr < Co, with ferric iron dominating at the oxidative end of the gradient and lead, chromium, and cobalt at the reductive side of the biomass.

In an effective regeneration (rehabilitation) of severely biofouled water wells, the expectation is that all the cationic accumulates would be dispersed to the groundwater if an effective treatment was applied. This means that a dispersed biomass recovered from a well would be likely to contain significant levels of accumulated cations dispersed during treatment and redevelopment. During the maturation of biofouling in water wells, one can expect sloughing (sheering) from the biomass will lead to what would appear to be random peaking of specific cations (e.g., arsenic) in the discharge. As maturation continues, the random peaks will transform into a plateau of higher concentrations as sheering advances to collapse of the biomass and causes plugging.

At the more oxidative side of the ORP gradient, it is not uncommon to observe changes in dominance within the alpha three consorm from iron to manganese or aluminum. For the Fe:Mn ratio it is normal for the ratio to remain stable at a particular site within a narrow range. The Fe:Al ratio can exhibit massive differences in neighboring bioconcretions that may be more reflective of local environmental conditions.

3, 10-21 (BPL) — Black plug layers in soils and turf grass golf greens are common. They are usually found as lateral layers of thick gelatinous biomass growing in the ground in the F region of the ORP in consorm cluster 3GF—a site of transitional movement from oxidative (above) to reductive (below). This commonly occurs at the static water level and can cause severe stress for the roots of plants including turf grass growing in soil or on greens. In golf greens, the symptoms in the turf grass shift through chlorotic states and wilting, leading to dieback, finally balding the green. The black plug layer consorm is competing with the plants for water, nutrients, and oxygen and essentially out-competing and killing the plants. Coring an infested green will show brown to black lateral streaks of slime where the black plug layer consorm dominates. Feeding solid fertilizer onto the green can exacerbate the condition by unintentionally also feeding the black plug layer. Control measures to prevent significant growth of black plug layer are aeration of the greens to drive down the ORP front and using liquid fertilizers that apply the nutrients directly to the leaves rather than the soil.

3, 10-27 (BBR)* — "Blueberries" are, in this context, spherical objects found littering the surface landscape of Mars and currently the subjects of conjecture. Such spherical objects are also found on Earth, primarily in bioconcretious growths such as rusticles. The bioconcretious biomass contains spherical structures usually 5 to 20 microns in diameter with high content of iron or another material with a waxy consistency. Iron-rich spheres are also found in pig iron deposits as terminal parts of the spent growth of ferric-rich bioconcretions. On Mars, the littering of "blueberries" may be connected to the collapse of bioconcretious growths that generated these spherical daughter products.

3, 03-16 (BWR) — Black water is water usually flows freely. Its black color is associated with iron sulfides and possibly carbonates. Black water normally originates from reductive conditions—high sulfate content and/or dissolved organic material. It is a product of hydrogen sulfide generation primarily by sulfate-reducing bacteria and secondarily via the degradation of sulphur-rich proteins present in iron-rich waters. The hydrogen sulfide reaction with any dissolved iron forms black iron sulfides. When excessive hydrogen sulfide is generated, rotten egg odor may be present.

---

* Speculative.

3, 19-03 (CRB) — Carbonates are plentiful on the surface of Earth, from mountain ranges such as the Dolomites to carbonate-rich geological formations. Many bacteria appear capable of generating carbonates as daughter products of metabolic activities, often in the presence of microbiologically generated electricity. Little is known of the mechanisms involved but carbonate-rich bioconcretions are commonly found at such sites as plugging wells and fouling cooling towers.

3, 05-27 (CGG) — Clogging is a geochemical term for a loss of transmissivity in water flows through porous and fractured structures moving toward water wells. Earlier thinking was that all production losses from water, oil, and gas wells were of geochemical origin. We now recognize that these production losses involve microbiologically influenced factors that are now considered forms of plugging (see also 2, 17-15, PLG and 3, 19-21, PLG in alpha groups 2 and 3).

3, 06-16 (CRS) — Corrosion is the deterioration of solid man-made fabricated materials as a result of aggressive chemical or microbiologically influenced events. Steel (iron alloy) materials are particularly at risk for microbiologically influenced corrosion (see also 2, 09-15 MIC). Corrosion can impact steel through lateral spalling and pitting that eventually lead to corrosion pits and subsequent perforation. Bacterial factors of great significance in corrosive processes are the generation of hydrogen sulfide (that initiates electrolytic corrosion) and acids. Acid-based corrosion is of two types. First, some bacteria can oxidize sulfides or sulfur to sulfuric acid to create radical drops in pH to as low as 0.5 to 3.0. Second, some bacteria reductively degrade organics to short-chained fatty acids; they can drop the pH range from 3.5 to 5.5.

3, 18-25 (CCR) — Concretions are crystallized structures dominated by ferric oxides, hydroxides, or carbonates. Natural concretions generally have very porous structures but synthetic concretions such as concrete tend to have smaller pores and denser structures. Generation of iron- and carbonate-based concretions involves the activities of various microorganisms including bacteria. Concretes remain subject to conjecture; most engineers believe that concrete is "cured" chemically; only a few believe that concrete is "grown." Growth in this case does not mean changes in dimension; it refers to structural changes to form and function of the concrete.

3. 15-23 (ECR) — Encrustations are formed by attached microbial biomass that also accumulates metals such as ferric iron. Carbonates may also be found within hardening structures. Bacterial communities live within and upon encrustations, usually at distinct sites that exhibit greater porosity and are often associated with channels formed within. As the encrustation matures, it maintains high porosity but becomes denser from accumulations of inorganic chemicals. For example, some encrustations can reach up to 80% or more gravimetrically as iron, at which time the bacteriological activity reduces (see also 3, 10-30, PGI).

3, 20-28 (FRD) — Ferric iron-rich deposits are generally formed at historic oxidation–reduction interfaces when a bacteriologically influenced biomass was active in the past. If the deposits do not become saturated with water, they may remain as geological artifacts with relatively little and/or diminishing bacterial activity. If the deposits become reductive in a water-saturated condition, the ferric forms of iron can become bacteriologically reduced to more soluble ferrous forms, leading to the dissipation of the ferric iron-rich deposit.

3, 22-18 (IPN) — Iron pans are laterally distributed ferric-rich layers within geological strata where at some time in the past an oxidation–reduction interface formed in concert with the static water level. Under such conditions, iron-related bacteria (IRB) initially generate a biomass along the lateral front that gradually bioaccumulates iron in ferric form under the more oxidative conditions found in the upper layers of the iron pans. The pans can be found at specific depths over many square kilometers. Under some circumstances, they become impermeable to water, causing permeating surface water and groundwater perching above the pan. Destruction of an iron pan can lead to a failure to pool water on top of the pan and development of more arid conditions resulting from loss of the perched water.

3, 18-14 (LSL) — A lateral slime layer arises from the growth of biomass along oxidation–reduction interfaces that are most common in lateral formations. These sites exhibit bioaccumulations of ferric iron (see iron pan). If the redox front moves as a result of changing climatic conditions, biomass generated at the redox front may also be expected to move (see also 3, 10-21, BPL).

3, 15-28 (MGN) — Manganese nodules develop on the ocean floor at the mud line. They are easily recognized as dense black spherical or plate-like objects containing high levels of manganese. These nodules consist of layers with very high bacteriological contents, indicating that the nodules may have grown as a result of bioaccumulative activities. Spherical manganese nodules often resemble the annular rings seen in the cortices of woody plants.

3, 18-19 (OCR) — Ochres are so named because they are deposits rich in ferric iron and commonly range in color from yellow to red to brown. The colors are generated by different environmental conditions and the forms of iron-related bacteria that dominate. Ochres usually form at oxidation–reduction interfaces and may interfere with water movements through formations by physically damming (clogging) the water flow or producing sufficient biomass to divert the water flow (plugging).

3, 10-30 (PGI) — Pig iron is one of the most common forms of the metal exploited to produce steels. Pig iron is found in large deposits in the crust of the Earth and reflects sites of significant oxidation–reduction gradients in ferrous-rich and organic-rich environments. At the shifting oxidation–reduction interface, intense bacterial activities would have included production of bioaccumulated ferric iron by iron-related bacteria. These deposits remain geologically intact because organics were insufficient to generate acidic conditions and the strata did not become reductive in a water-saturated state. Chemically PGI is identical to matured, spent, collapsed rusticles (see also 3, 19-26, RST) scattered around sunken steel-clad shipwrecks.

3, 03-19 (PFR) — Perforation is a terminal event that occurs when a layer of a material such as steel is subjected to microbiologically influenced corrosive processes. Perforation normally begins as pitting that may pierce the material or involve the formation of lateral cavitation (dishing, gouging) that finally leads to the more generalized perforation event. Perforation is a major factor in the failures of steel-clad ships (requiring major repairs or causing the scrapping or sinking of the vessel), water tanks and distribution systems (leaks), gas lines (flare-off events), and oil lines (incurring major environmental clean-up costs).

3, 06-24 (PTG) — Pitting of steel commonly begins with the growth of a biofilm on the surface. The site of such growths may be environmentally established but may involve electrical charges on the surfaces of conductive materials. Pitting is more likely to begin at anodically charged sites that attract bacterial attachment. Steel usually contain phosphorus and sulfur—both are desirable macronutrients for a growing bacterial biomass. Bacteria will attempt to "mine" these nutrients from the steel by generating fatty acids (causing acidulolytic corrosion) or hydrogen sulfide (causing electrolytic corrosion). Pitting in steel forms as a vertical progressive dissolution that eventually leads to perforation (see also 3, 03-19, PFR).

3, 19-21 (PLG) — Plugging is a condition in which water flows through a porous medium (e.g., sand) or fractures (e.g., rock) and is infested by the growth of a biomass along the flow path. As the biomass grows, the first effect is diversion of groundwater flow around the plugging biomass followed by a complete stop of flow through the infested region. Plugging can be caused by alpha two or alpha three consorms but the nature of the alpha three (higher inorganic content in bioconcretions) makes its biomass much less prone to sheering effects from water flow. Regeneration of such plugs becomes more challenging when the plugging biomass has high ferric or carbonate content (see also 2, 19-21, PLG).

3, 19-26 (RST) — Rusticles are forms of growing biomass that often dominate the surfaces of steel, particularly when submerged in seawater under oxidative conditions. Rusticles are bacteriologically influenced concretions that develop complex structures with high porosities and usually high ferric iron contents. When environmental conditions are favorable, iron may be replaced by a more dominant metal such as aluminum. Rusticles terminate as forms of pig iron.

3, 10-10 (SBL) — Sand boils are bacteriologically mediated events in sand structures such as levees in which biomass infestations have occurred. The passage of water through the sand may be impeded by the biomass. When the back pressures become too severe, the lines of least resistance come into play and the water pressure is relieved along those pathways, avoiding the active plugging biomass. Sand along the pathway becomes unstable and water appears to bubble up (boil) through the sand following the lines of least hydrostatic resistance to flow.

3, 15-17 (TCL) — Tubercles are close cousins of nodules but differ in two ways: (1) the tubercle tends to have an ellipsoid base to the dome, and (2) it will show evidence of eruptions along the top of the dome. These eruptions are generally very slow to form and become fossilized into the peak of the tubercle structure as bioaccumulated ferric iron.

4.5 ALPHA FOUR: CARBON-REDUCING CONSORMS (FPL 4, 06-27)

Conjecturally, alpha four bacteria dominate in reductive environments, stripping organic materials (synthesized primarily in the surface biosphere) from water-saturated regions of the Earth's crust. Organic materials generated under oxidative conditions involve carbon, hydrogen, oxygen, nitrogen, phosphorus, sulfur, iron, manganese, and a range of micronutrients. These organics are drawn downward via diffusion and gravity conditions shift across the oxidative–reductive interface. Along this gradient to increasingly negative ORP values, the activities of alpha four consorms dominate, stripping off the oxygen, nitrogen, phosphorus, sulfur, iron, and manganese along with the trace elements, to leave hydrocarbons (C_xH_y) or carbon. Daughter products of the activities of alpha four consorms are hydrocarbons or elemental carbon. The classification of the alpha four consorms is therefore based on the form of daughter products they generate. Typical calcitrant daughter products include petroleum hydrocarbons, natural gas, coal, tar sands, and crude oils.

4.5.1 Principal Alpha Four Consormial Activities

Alpha four consorms are differentiated primarily by daughter products generated from their reductive metabolic activities. The reductive degradation of organics down to hydrocarbons appears to involve common intermediaries, shorter chained fatty acids. The acids are generated by many microbiologically driven fermentative processes and they may provide primary organic carbon sources for the alpha four consorms.

The principal daughter products of interest arising from alpha four consorm activities are all hydrocarbons (C_xH_y) and the differentiation of these consorms is based on the form in which these products are generated. The forms range from extended carbon chains as polymers typically found in petroleum hydrocarbons to the shorter chained forms generated in natural gas. Further consorms in the alpha four group reduce organics to both C_xH_y and elemental carbon from which the hydrogen content is also stripped. These elemental carbon-dominated forms are found in coal-based products.

4, 05-25 (BAP) — Black asphaltene-rich plugs are known as black "goop" and commonly found in oil transmission lines and oil wells extracting heavier grades of crude. These plugs form when bacterial consorms within the oil "mine" water from the oil and then protect their newly acquired (bound) water in asphaltene-rich coatings. The plugs generally attach to pipe walls and can cause losses in flow that result in elevated energy costs to pump the oil. Such plugs can form natural sites for initiation of microbiologically influenced corrosion that may compromise pipelines when perforations in the pipe walls cause leakages.

4, 04-20 (CDR) — Carbon dioxide reducers (CDR) are natural bacterial consorms that trigger the phenomenon by which carbon dioxide is deliberately entrapped and reduced within geological formations. This process is also known as carbon capture storage (CCS) or carbon capture sequestration (CCS). Little is known of the microbiological impacts triggered by the entrapment and leading to bacteriological reduction of carbon dioxide. It is, however, relatively certain that the carbon dioxide entering a descending flow into the crust will be bacteriologically reduced to petroleum hydrocarbons, methane gas, and coal, depending on local environmental conditions. A further risk is significant plugging (see also 2, 19-21 and 3, 19-21, PLG for more information on plugging) that may divert flows.

4, 05-17 (COL) — Coal is a matured geological product in which most of the organic material is reduced from structurally complex forms to dominance by elemental carbon generated by microbiological activities. Laboratory studies indicate that coal can form particularly at anodically charged sites. Black coal has a high potential energy output (by combustion), may be nutrient-rich, and retain some level of bacteriological activity.

4, 06-22 (BNG) — Biogenic natural gas is generated under very reductive conditions. Some bacteria can generate methane as a principal metabolic daughter product. They are known as methane-producing bacteria (MPB). Carbon sources reduced by these bacteria include shorter chained fatty acids and carbon dioxide. Generally, shallower formations of natural gas deposits are dominated by methane generated by MPB under very reductive conditions at temperatures below 50°C. Natural gas (primarily methane or CH_4) is a major source of energy. It can be produced biogenically by anaerobic bacteria, but methane from deeper formations has been traditionally considered a product of high temperature geochemical (thermogenic) reactions. Biogenic and thermogenic forms of methane are distinguished by the ratio generated by the analysis of the two stable carbon isotopes, ^{12}C and ^{13}C. Only 1 in about 39 carbon atoms is ^{13}C in methane generated in a surface environment. Thermogenic methane has been linked to -40 to -20% ranges for ^{13}C in CH_4. It is uncertain whether some of the thermogenic natural gases determined by fitting into this carbon isotopic ratio are not simply from more deeply generated biogenic methane at more extreme environments of higher temperatures and pressures in saturated groundwaters.

4, 03-13 (BSR) — Black smokers are found at deep-ocean sites along fault lines (e.g., the Mid-Atlantic ridge) where hot sites with high sulfate content waters saturated with methane are released through vents and seeps. This activity creates a complex bacterial community including the SRB that reduce sulfates to hydrogen sulfide (the black color comes from sulfides) and methane-consuming bacteria that dominate when conditions are oxidative (further up the vent and out into the water). The biomass thus created around the vents of black smokers becomes feeding stock for many animals that grow near the oxidative sides of vents in cooler waters.

4, 01-28 (OIL) — Oil falls into a broad group of total petroleum hydrocarbons (TPHs)—a large family of several hundred chemical compounds that form a major part of crude oil. Oils are made under very reductive conditions. Oil is commonly dominated by C_xH_y with lengths of 9 to 50 carbons. Shorter chain lengths tend to be dominated by gaseous or volatile forms of hydrocarbons. Because of the number of different chemicals in crude oil and other petroleum products, it is not practical to measure each chemical separately. Common practice therefore is to measure the total TPH at a given site. Local environmental variations from reservoir to reservoir create considerable natural variation. Some important chemicals found in TPH are hexane, gasoline, jet fuel, mineral oil, benzene, toluene, xylene, naphthalene, and fluorene.

4, 07-19 (PET) — Peat originates from organics deposited under water-saturated reductive conditions. Fermentation arising from releases of fatty acids creates a chemistry that restrains and then totally halts microbiological activity. Under these conditions (such as in a peat bog), degradative processes are halted. Bodies found in peat bogs still have preserved features and forms due to the lack of degradation of body tissues and clothing. Peat has high carbon content, can burn under low moisture conditions and, like lignites, is considered a form of coal. After ignition by a heat source (e.g., a wildfire penetrating the subsurface) a peat fire can smolder and burn undetected for very long periods (months, years, even centuries) propagating in a creeping fashion through the underground peat layer. Peat fires are emerging as global threats posing significant economic, social, and ecological impacts.

4.6   ALPHA FIVE: CARBON-OXIDIZING CONSORMS (FPL 5, 13-07)

All components of a living system require some level of interchange to allow a stable and integrated cycle of life to proceed. This occurs at the organismal, communal, consormial, and even at environmental levels. The first four consormial groups (alpha one, two, three, and four) all participate in the gradual reduction of carbon and carbon dioxide to more reductive forms, whether biomass, bioconcretions, or hydrocarbons. These consorms follow stages within the gradual movement of the environment from oxidative dominance to one that is progressively reductive. This sequence of biological activities terminates in pooling of the reduced forms of carbon (primarily hydrocarbons) in the most reductive phases in the water-saturated geosphere.

While this reduced carbon may pool for some time as coals, oils, gases and tars, it is also subjected to various physical gradients at the pooling sites including gravity, electrical forces, temperature, pressure (including sudden venting releases), and geological instability. These reduced forms are carbon can move upward toward the oxidative surface biosphere or downward into the extremely hot and pressurized magma interfaces. The alpha five consorms play major roles in the entrapment and oxidation of these reduced carbon compounds. Many of these entrapment and carbon oxidation activities are already normal functions of the alpha three consorms in the ORP transitional regions, the alpha two consorms in the oxidative regions, and the alpha one groups where the reduced carbon enters a gaseous phase and water is restricted into dispersed or aerosol droplet forms. These bacterial consorms have already been established. Alpha five consorms are restricted to those that cause some accumulation of reduced forms of carbon, usually at the oxidative–reductive interfaces as the environment moves upward from reductive to oxidative.

4.6.1   Principal Alpha Five Consormial Activities

The limitation defining alpha five consorms (ability to accumulate and oxidize reduced carbon compounds) means that few locations are available for this activity to occur in a definable manner. One suitable site would be where volatile and gaseous hydrocarbons move upward toward an oxidative–reductive interface where a triggered response accumulates hydrocarbons subjected to controlled degradation as sources of energy and carbon. Only one consorm clearly fits this category—the one involved in the functioning of clathrates, more commonly known as gas hydrates, and the largest global resource for natural gas.

5, 15-10 (GHY) — Gas hydrates are complex methane–ice structures that form in deep-ocean environments where the temperatures are in the 3 to 7°C range. Gas hydrates are also known as clathrates and tend to form under reductive conditions at sites of periodic releases of natural gases. They can entrap and cluster methane around water ice in a ratio of 8:1 and occur along continental shelves and under frozen tundras. Gas hydrates may represent the most significant untapped sources of natural gas on Earth. The bacterial component is present in at least two distinctive phases: (1) reductive conditions in which the bacteria generate ice clusters that, through the generation of bacterially influenced surfactants, accumulate methane molecules at a ratio of 8:1 to every water molecule; and (2) under oxidative conditions allowing controlled utilization of methane to provide energy and carbon for growth.

Each clathrate can therefore be considered biphasic—having two distinct functions of reductive accumulation followed by oxidative degradation of the methane. Studies of gas hydrate consorms are ongoing but have been severely hampered by low temperatures and high environmental risk that the consorms may be converted to functionally unstable communities.

4.7 ALPHA SIX: HYPERBARIC DISPERSED BIONUCLEATING CONSORMS (FPL 6, 01-03)

Alpha six consorms are speculated to be associated with the extreme environment between the very hot magma that appears to be water-free and the water-saturated geological media forming the crustal elements of the planet. Water moving downward into this very hot pressurized interfacial zone between crust and magma may be expected to vaporize as steam. The steam may move downward into the magma, form an envelope between the magma and the crustal elements, or move upward into the crust where condensation into liquid water can again occur. The latter event could trigger a cycling of water from the liquid state (in the crust) to a gaseous state (in the envelope above the magma). This would mean that water cycle from the gaseous to a liquid phase at very high pressures and temperatures. Water in the liquid phase has the potential for bacteriological activity. This means that life in the form of bacterial activity could occur in the liquid phases of the water associated with the magma crustal interface. Based on the length of time these events required during evolution and the adaptation to changes in diverse environments near the surface biosphere, the premise is that life forms adapted to those conditions. The alpha six bacterial consorms remain speculative at present and the challenge is to construct conditions that would validate or deny their existence.

5 Environmental Dynamics of Bacterial Consorms

5.1 INTRODUCTION

Before explaining how to recognize the various bacterial consorms defined via the alpha classification scheme (first covered in Chapter 3), it is important to gain some understanding of the concepts used. The traditional need was to ensure that an individual bacterial culture was "pure" (not a contaminated mixture of bacterial strains). The first challenge in defining bacterial consorms must be to adapt relationships to the definition of the term "pure" to define dynamic communities of bacteria that are, by their very nature, mixtures of bacterial strains. Defining a bacterial consorm involves descriptive observations of the growing biomass within its natural environment along with details of the nature and form of the environment within which it flourishes. Note that a number of implicit errors inevitably occur as soon as samples are taken from the biomass:

1. Selective isolation of the part of the consorm removed.
2. Disruption of the natural environment within which the consorm was active, causing trauma that may lead to radical changes in the bacteriological composition of the biomass.
3. Post-sampling handling of the selected biomass causing significant changes in the activities of the component bacterial communities.
4. Any selective cultural technique is likely to cause further distortions within the surviving bacterial communities.

These potential errors mean laboratory investigation of a bacterial consorm is likely to give a distorted (but replicable) scientific answer that may or may not be reflective of the true nature of the consorm. Thus, the identification of bacterial consorms centers more on the original site (habitat) within which the biomass was first identified and described. Bacteriological investigation of such consortial biomasses is challenging because of the potential for major distortions in determining the nature of the consorm. The challenge extends to the practical protocols be used to recognize and categorize the various bacteriologically dominated consorms. This book represents a first attempt to "lump" bacteria into consorms in a manner that will have significance in the applied world where corrosion, plugging, and process efficiencies are significant concerns. Practical analyses of consorms utilizing chemical and cultural techniques are described in Chapters 8 and 9.

5.2 DEFINING BACTERIOLOGICALLY DOMINATED CONSORMS

In practice, one of the first principles for a bacteriologist wishing to identify a particular bacteriologically dominated consorm would be to achieve a level of "comfortable" identification of the consorm. Splitters (reductionists) always want to work with pure cultures. "Pure culture" may be loosely defined as any culture of bacteria in which all the component cells have been grown from a single cell or all the cells bear a completely common set of characteristics. In other words, the growth (culture) develops from a common identical stock of bacteria. Ideally, the culture should consist of bacterial cells that are all clones of each other (are identical) and exhibit complete genetic homogeneity. The definition of "consorm" invokes a dynamic state within a biomass. For this purpose, the definition is:

> Consorms are created within natural environments when a dynamic integrated, bacteriologically influenced community structure has developed. This consortial structure involves as active participants all the different bacteria that are present and each performs a unique role in support of the consorm. Such activities are not primarily driven by a single dominant species, but by the coordinated activities associated with the various members of the integrated bacterial communities. The composition of consorm is driven by the ability to exploit sources of energy, nutrients, and water to drive the community. The nature of the consorm can best be determined by the structures, forms, and functions of the bacteriological communities that play functional roles associated with the consorm.

The complex interactive structural, biochemical, and physical nature of a specific environment is implicit in defining the natural generation of a consorm. That concept imposes basic scientific challenges in the culture, identification, and classification techniques. Traditional protocols require that growth be definable by its characteristics for an isolated strain from a culture, followed by intense biochemical and genetic characterization if needed. These characteristics are normally more discernible based on observations of a selected colony growing on or within an agar-based culture medium rather than in a liquid medium (broth). Agar media often contain chemical and physical agents that aid in the (primarily visual) selection of an appropriate colony for use in the identification protocol but also limit the spectrum of bacteria cultured. In the identification of a consorm, any cultivation in a selective manner (such as on an agar medium) would distort the findings. While such distortions may be replicable, they would not allow a comprehensive determination of the bacteriological nature of the consorm.

Significant challenges surround the culture of consorms. Unlike pure cultures, the component bacteria within a consorm constantly change in a dynamic manner to respond to environmental conditions. Even within the consorm, dynamic shifting of environmental conditions may cause component bacterial species within the consorm to cluster within communities that would then contribute separately to the functioning of the consorm as a whole. During culturing of a consorm, changes resulting from the technique may cause significant changes in the species present in the consorm. Such constant adjustments in the nature of the bacterial composition of the active consorm may arise from interactions among the communities and render culturing a challenging activity. As a result, the product of the culture is likely to vary in a manner that would cause concern in developing scientific principles for the classification and functional enumeration of consorms.

Microbiology practice during the past 150 years focused on the need to extract a single species from a consorm or sample and concentrate on that species as a representative of the bacteria in the sampled site. This reductionist approach led to dominance of the "splitter" philosophy in which the refined pure culture of a single bacterial strain is considered to provide replicable conditions that allow scientific validation. Examining a consorm scientifically is challenging because all the bacteria within various communities interplay at several levels, leading to a potential inability to replicate data from the consorm in a manner typical of "pure" cultures. In the lumper technique for evaluating consorms, emphasis moves from the cellular and subcellular levels investigated by splitters to the environmental habitat and communal characteristics researched by lumpers. This creates a fundamental challenge: moving from the definition of a single culturable species to the analysis of a dynamic biomass treated as a whole.

Any attempt to culture consorms creates challenges due to changes in dominance that may occur. Defining consorms therefore uses a set of parameters different from the technique of identifying individual bacterial strains primarily by biochemical and genetic characterization. Parameters for the identification of consorms concentrate on the nature, structure, and form of the consorm along with characterization of the environment in which the consorm grows. Implicit in lumping classification is the principle that the consorm is identified by common features, not by characteristics of single bacterial communities within the consorm despite the assumption that bacterial communities in a consorm are too dynamic to allow adequate precision or provide significant critical parameters to recognize the consorm as a definable entity.

Unlike the microbiology method of culturing followed by biochemical determination of a specific type of isolated bacterium, identification of consorms involves no cultural practices; it requires observations relating to the intact biomass and habitat in which the consorm grows. Entering a consormial biomass for sampling purposes inevitably causes disruption that could be fatal to the normal functioning of the consorm. A researcher should consider a consorm a complex living organism like a plant or animal rather than simply treating it as a collection of bacterial cells in a haphazard arrangement. Just as a physical intrusion into a plant or animal body at least minimally causes tissue damage, intrusion into a consorm disrupts the delicate harmony established among its component bacterial communities.

5.3 CATEGORIZATION OF CONSORMS

Much of the identification of a specific consorm should be achieved by observations relating to the environmental location of the biomass and its form, function, and appearance. Replicating a consormial biomass in a laboratory presents challenges: sampling, transportation, precise duplication of critical environmental factors (e.g., oxidation–reduction potential, water chemistry and state, temperature, pH, critical nutrients, nature of surfaces and voids within or upon which the consormial biomass will attach and grow). Another need is to ensure that all communities active within the consorm are present and able to integrate into a single common functional biomass. In summary, in defining a bacteriologically dominated consorm, the major identifiable features relate to the in situ nature of the biomass and include:

1. Location within the environment supporting the activity of the consorm
2. Primary colors generated within the biomass
3. Textures and porosity
4. Inorganic and organic chemistry of the biomass
5. Water form and content
6. Odors emanating from the consorm
7. Determination of bacterial communities recovered from the biomass
8. Time-scaled maturation

The nature of a consorm is to be distributed widely within an environment. It is unrealistic to expect to recognize the whole biomass associated with the consorm; investigation should focus on some part of the consorm that forms an observable component. A parallel may be drawn from the fruiting caps of *Agaricus campestris* mushrooms. The biomass of the fruiting caps garners much of the attention although the caps form only 5% of the total biomass. The large unrecognized biomass consists of mycelial threads permeating the supportive porous media. Thus a consorm should not be determined only by what is obvious; it should be investigated in relation to its biomass and interconnection with the surrounding environment.

5.3.1 LOCATION WITHIN ENVIRONMENT

Activities within a consorm must be products of the nature of the precise environment of the consorm. A number of factors affect the positioning of a consorm biomass within a given environment. Generally, bacteriologically dominated consorms tend to have water content (in the void or on the surface volume) exceeding 80%. Under oxidative conditions, water content below 80% in a biomass causes competition with fungi that have far better tolerance of low water viscosity conditions. Under reductive conditions, reduced available water volume below saturation may be caused by gas vacuoles rather than direct competition from other organisms. Also, reductive conditions cause the bacteria to adapt to anoxic conditions, then dominate and exclude all other organisms.

Critical in the evaluation of location is the oxidation–reduction potential (ORP) that signals biological competition for bacteria on the oxidative side but little competition on the reductive side. At the oxidative–reductive interface, conditions move from oxidative to reductive. It serves as a focal site where bacteriologically dominated consormial biomass growths occur. The front occurs at ORP values of +50 to –100 millivolts (mv) and provides certain advantages for a consorm:

1. Protection from competition created by other organisms that can only function in oxidative conditions
2. Ongoing access to oxygen although at critically minimal levels for bacterial aerobic functions unless electrolysis becomes a significant factor

3. Access to reduced soluble chemicals moving from the reduced environment into the ORP front where nutrients are assimilated and toxic chemicals accumulated within the biomass
4. Saturation with water, eliminating potential for an air:water interface where fungal (mold) growth could dominate

Bacterial consorms generally attached to two primary surface forms: (1) exposed surfaces within fractures or voids of porous media; the probability is that the consormial biomass would terminate by filling the fractures or voids, thus creating a plug; (2) the surfaces may be farther apart, e.g., in a fracture or on an exposed surface. Under these circumstances, the formed biomass may resemble a coating or slime. A coating is generated when the infested surface generates an even coat that may be dominated by bioaccumulated chemicals (e.g., ferric iron, carbonates); when the surface is coated with a viscid layer dominated by biofilms, the result is a slime.

Observation of a bacterial consorm is rendered more challenging particularly when it constitutes a biomass plugging the voids of a porous medium. It is difficult to view a plugging consorm in a porous medium without disturbing the medium. Investigating a fracture would involve coring rocks to reach the plugging consorm. Usually, a consorm causing a physical occlusion of water pathways may be measured by changes in the geohydrological characteristics of water that has passed through the infested zone. Generally the first observations relate to the oxidative side of the consorm-infested region and are influenced by bacteria functioning on the more oxidative side of the consormial biomass.

5.3.2 GENERATED COLORS

Color is a valuable characteristic of a consorm. It is usually stable but may change as a biomass ages. Consorms present a greater variety in colors under oxidative rather than reductive conditions. This means that the first color observed may be atypical of the consormial biomass as an integrated whole structure under more reductive environmental conditions. The color observed may be transitional. Even so, color represents represent a valuable insight into the nature of bacteria within a consorm and the environment in which the consorm is active. Color categorization therefore has limited value due to these variabilities. The various colors observed in bacteriologically influenced consorms are described below.

Black is commonly observed in consormial biomasses growing under reductive conditions, particularly where the ORP value lies between -20 and -150 mv. The most common cause of a jet black color is the presence of iron sulfides or carbonates. Sulfides are products of the bacteriological generation of hydrogen sulfide under reductive conditions; they react with iron cations to form iron sulfides. Some iron-related bacteria (IRB) in consorms can generate carbon dioxide that also reacts with iron cations to generate black iron carbonates under reductive conditions; this type of black growth is common also under reductive water-saturated conditions.

The formation of iron sulfides generates hydrogen sulfide by reducing sulfates or biodegrading sulfur amino acids. The resulting black consormial growth appears under very corrosive conditions such as in the deeper areas of steel pitting, but may also be observed in lateral black slime layers active at the oxidative–reductive inter-face. Consormial growths may also generate black slimes in relatively reductive regions. The slimes become visible as the growths move into oxidative regions where air is present. The black color may become fainter as the sulfides are bacteriologi-cally oxidized to sulfates and sulfuric acid. Black plug layers may occur in soils and golf course greens if the static water level is stable and near the surface of the soil or green. Plug layers may be sufficiently tight to prevent the downward movement of water and compete with plant and grass roots causing stress, that can lead to dieback directly over the growths.

Grey color appears in consorm growths for a variety of reasons. Grey displays a range of shades from light to dark. Grey is not generated by single bacterial commu-nity; it commonly reflects a number of concomitant events including the generation of carbonates (white) and iron sulfides (black) within a biomass. Other factors are bacterial communities that bear a white (common) or black (rare) pigment. Thus, grey color has little significance because it tends to be relatively unstable and reflects several bacteriologically influenced activities.

White consormial growths occur mostly under oxidative conditions; consorm-ial pigments are generated via the formation of carbonates and/or white pigments produced by some of the bacteria within the consorm. If carbonates (commonly calcium carbonates) are formed, the growth will harden into a solid structure that remains white even under reductive conditions. If the pigments are generated by some bacteria within the consorm, the white color will "shade out" to another colour (e.g., cream, yellow) or become transparent.

Cream is a very pale yellow or off-white color found in consormial growths and usually reflects mixed bacterial communities that include strains generating white or yellow pigments. Natural yellow natural pigments are often classified as flavines produced by certain bacteria. Cream-colored biomass tends to occur under oxida-tive conditions and becomes unstable as the bacterial communities in the consorm mature.

Orange is composed of reddish yellow hues. Yellow hues may be generated by bacterial flavine-like pigments; red may be generated by ferric compounds when accumulated in a biomass at low concentrations. At higher concentrations, the ferric forms generate brown hues. Red hues are more associated with bacterial pigments, e.g., those produced by enteric bacteria and anoxygenic sulfur bacteria. While enteric bacteria may be active in oxidative conditions, anoxygenic sulfur types appear only under reductive conditions where some light is available. Orange is not likely to arise from interactions of enteric bacteria with bacteria that gener-ate flavines.

Red color is relatively rare and may range in hue from pale pink to dark red and even purple. Under reductive conditions with low levels of light, red colors are generated by anoxygenic sulfur bacteria. Under oxidative conditions, red pigments are generally only by a limited range of bacterial species; genus Serratia are most common.

Light brown represents a mixture of red, yellow, and black. While bacteria may generate such pigments, this color is primarily caused by ferric iron compounds in the consormial biomass—usually by the ferric oxide and hydroxide complexes accumulating within the biomass to provide structural support. Light brown arises from bioaccumulated ferric compounds that represent 20 to 40% of the total dried weight of a consorm. If the concentration exceeds 40%, the color darkens.

Dark brown is another mixture of red, yellow and black (see paragraph above for details of the origins of brown colors). In more mature, iron-rich bacteriologically dominated consorms, the ferric iron concentration rises from 40% as high as 95% of dried weight of the bioconcretion. As the ferric content increases, the brown shade darkens. The consorm biomass within the bioconcretion may also accumulate other metals; two that may become dominant in the chemistry of the biomass are aluminum and manganese. Aluminum can cause bioconcretion color to lighten to the point of becoming white; manganese darkens and the growth appears black.

Violet (bluish-purple) appears opposite red in the spectrum. Very few bacteria develop purple pigments. Purple is usually generated by a consormial biomass in an organic-rich oxidative environment. The color is most intense where mucoid growth interfaces directly with air. Often the background of the underpinning growth is white and the contrast may take on a sharp edge between the violet and white. Two bacterial genera commonly associated with purple pigments are Chromobacterium and Janthinobacterium.

Yellow pigments are generated by many bacteria. Yellow lies between green and orange in the spectrum and commonly indicates various flavine-like compounds that generally intensify with maturation of the growth and in the presence of light. Yellow pigments are common in the bacterial kingdom; in a biomass, they may result from the dominance by certain bacteria.

Fluorescent green color emanates from consormial growths that are oxidative when exposed to ultraviolet light. The color extends through the greenish-yellow parts of the spectrum and appears to glow when exposed to ultraviolet light. A number of Pseudomonas species emit a greenish glow that may last 2 days or more during maturation of a biomass and may remain several weeks.

Fluorescent blue color is emitted by consormial growths that are oxidative when exposed to ultraviolet light and appear to glow. Several Pseudomonas species produce a transient pale blue glow lasting a couple days or more during maturation of the biomass; in most cases, the glow disappears within a week.

Bioluminescence is a phenomenon by which a bacteriologically generated consorm generates light that is usually emitted in the blue part of the spectrum. Generation of light by organisms is an energy-expensive process involving the utilization of high energy reserves tied up in adenosine triphosphate. The energy is released by the luciferase catalyst that generates light. The firefly is a classic example of an organism that generates light. Marine life also generates bioluminescence, for example, many deep ocean-dwelling fish carry light-generating organelles that consist of entrapped bioluminescent bacterial consorms. A not-so-well-known example occurs in the deep ocean layers from 400 to 1000 meters down. Continuous flickering of bioluminescent blue light makes the ocean look like a giant suburban development extending as far as the eye can see. Generally bacteria-producing bioluminescence are classified into genus Photobacterium. A bioconcretion that glows with a blue light may indicate the presence of bioluminescent bacteria.

5.3.3 TEXTURE AND POROSITY

Texture is based on the forms of surface structures and porosity relates to the number of voids (spaces) within a consormial biomass. The two approaches to determining texture are: (1) visual examination of the growth; and (2) determination of the nature of the biomass, particularly its structure and strength. These two attributes will be addressed separately below but they are intimately linked. Porosity measures the volume of water in a biomass in relation to the total volume of growth.

Visual inspection of a consormial biomass has two limitations. The first is direct evidence associated with porous or aqueous media. The second is evidence created by disturbing the supporting medium in which a biomass is active. Clearly "disturbance" may change the nature of a revealed biomass so that it becomes atypical of its natural undisturbed state.

Examination of a primary (visible) consormial biomass is relatively easy, particularly if it is in or attached to a liquid medium or solid substrate. A biomass may be suspended as a cloud-like entity in water or may form at the interface of water with a solid substrate. The suspension may be at the oxidative–reductive interface or set directly on solid media (clays, sands, and muds) beneath the water column. A glistening surface indicates a strong probability of significant water content; low reflectivity means higher solid content. Generally a biomass growing under these conditions is likely to accumulate various chemicals that have no immediate use. The bioaccumulation becomes apparent as the density of the biomass elevates. Such accumulations where growth is at the oxidation–reduction interface are exaggerated by the active oxidation of various cations, e.g., the conversion of ferrous iron to the less soluble ferric forms that contribute to the density increases.

Textures may also be affected by the casual bioaccumulation of materials suspended in upstream water. Sands, silts, and clays may be passively accumulated in a growing biomass. They appear to play no active role in the biomass but occupy significant volumes in voids. One effect is that when a treatment is applied to shock, disrupt, and disperse the biomass, the accumulates are released into the downstream water flow. The sudden appearance of sands, silts, or clays downstream may be indicative of a release of passive accumulates by a dispersing consormial biomass.

In addition to examining texture visually, it is also important to determine the nature of the materials. In the field, the simplest test to composition of texture is to rub the biomass between fingers and thumb in a rotatory manner. If the biomass sample is dominated by sands, the hard granular nature of individual sand particles becomes evident. Silts tend to display minimal granular structure and slide easily between finger and thumb without obvious smearing. Clays smear and make a material slippery. If the accumulates were generated actively, the dominant materials are likely to be ferric forms of iron that produce a granular texture and pronounced smearing due to (commonly) yellows, oranges, and browns that dominate the smear.

Most consormial biomass structures include many actively accumulated metallic cations that tend to be positioned at different ORP values. In the successful dispersion of a biomass, such cations will appear in downstream waters at different times—the more oxidatively accumulated cations first, followed by cations accumulated under progressively more reductive conditions in sequence. These changes in chemistry are discussed in Section 5.3.4 below.

These changes in ORP values in the environment as the biomass moves from oxidative to reductive conditions can cause structural changes in the biomass caused by domination of different bacterial communities. These shifts in dominance primarily from aerobic (oxidative) to anaerobic (reductive) may affect the nature of the dominant bacteria in the biomass. Sampling such a dynamic consortial form of bacterial community creates challenge in interpreting the nature, form, and function of the biomass. Because of these additional challenges, it is often easier to simply categorize biomass growth by the environment within which it grows, using simple descriptors without detailed bacteriological analysis.

The structural nature of a bacteriological consormial biomass is very different from the structures of plants and animals. Higher organisms are composed of differentiated tissue cells plus microbial cells that play a parasitic role within a host organism. A bacteriologically dominated biomass always includes a considerable percentage of inorganic materials structured outside the cells but within the biomass. As a result, the texture of a bacterial consorm may be dominated by these extracellular materials. In young consorms these materials are dominated by water; as a consorm ages, the dominant materials outside cells move from water to inorganic crystalline structures such as ferric iron oxides and hydroxides or carbonates.

The net effect of this ongoing accumulation of inorganic materials within a consorm is an increase in density and changes in structural appearance. These changes commonly occur in the transition from a mucoid water-rich biomass to one dominated by complex crystalline structures. Structures on the oxidative sides of such growths may appear hardened with flake-like or concretious surfaces. On the reductive sides that may be limited to some parts of the consorm, water retention may exhibit less development of oxidized products such as ferric iron compounds. Perhaps the only general statement that may apply is that water content declines while the inorganic structures increase within the total biomass of a consorm.

Porosity is defined by the aging of a consormial biomass. During aging, the ongoing bioaccumulation of surplus chemicals beyond basic metabolic needs leads to an increase in density and decrease of water content. Certain bioaccumulated chemicals (such as carbonates and ferric iron) become structurally significant, causing the diversion of water from the smaller voids into the growing water passages through the biomass. The two effects of this maturation are: (1) the total volume of water within a given volume of biomass declines; and (2) bacterial communities within the consorm tend to be isolated because they occupy less available water volume and thus separate from each other.

This creates a challenge for identifying bacteriological communities within a given volume of biomass since increasing volume is required to determine the presence, activities, and reactions of these communities within a sample. In a young bioconcretion, for example, volumes as small as 10 ml allow recovery of the various communities; increased volumes (100, 1,000, and eventually 10,000 ml) are needed as maturation progresses. In the case of mature iron-rich bioconcretions such as rusticles on sunken steel shipwrecks, it is easier to track water paths through specimen materials and take samples along those paths. Low porosity or water content means more difficulty in recovering active bacterial communities within a consormial biomass as distinctly separate clusters.

Cultural enrichment techniques must be applied to detect the activities and reactions of specific bacterial communities. The BART™ (Biological Activity Reaction Test) system (Droycon Bioconcepts Inc., Regina, Canada) described in Chapter 9 allows differential culturing of specific bacterial communities based on the natures of the oxidation–reduction interfaces created as selective chemical media diffuse up into a sample being tested. This system ensures that bacterial communities within a solid or liquid sample have time to adjust to and become active in the selected segment of the ORP gradient. This technique for selective enrichment may be applied to consormial biomass structures of varying maturities.

5.3.4 Inorganic and Organic Chemistry

It is normal to obtain a chemical signature from a given sample of a consortial biomass to investigate bacteriologically influenced activities and reactions. Inorganic determinations have become more precise and metallic cationic analysis is the most refined test. Less refined is the general analysis of organic chemicals made more challenging by the vast diversity of organic compounds. Water samples are routinely subjected to extensive and expensive inorganic analyses while organic analysis is often neglected and bacteriological analysis is limited to determining the presence or absence of coliform bacteria.

In reality, significant bacteriological activities and reactions can be ascertained globally in a solid or liquid sample by assessing total phosphorus, nitrogen, organic carbon, and adenosine triphosphate (ATP) as the primary energy driver (see Chapter 8). Attempting to increase precision by determining dissolved forms such as total dissolved organic carbon can create difficulties in the presence of an active bacteriological population. This population may consist of dispersed cell contents and extracellular detritus that may create significant bacteriologically influenced chemical activity. Each major nutrient (organic carbon, phosphorus. and nitrogen) follows a unique biochemical pathway that challenges interpretation except when the only assessment is total amount of chemical and even those results exhibit major differences.

Organic carbon, for example, assumes a multitude of molecular forms exhibiting different chemical and physical reactivities in dissolved, biocolloidal, and suspended particulate states. Where biochemical activity occurs in a solid or liquid sample, the most reliable parameter may be total organic carbon (TOC) although TOC determination represents only an estimate for a particular moment in time. An increase in TOC over time means that inorganic forms of carbon are oxidizing into organic forms via biological activity. TOC may also decline over time if some of organic carbon is degraded to inorganic forms including carbon dioxide.

Phosphorus has a very different mass balance because it exists in four primary forms that function very dynamically inside and outside a biomass: (1) soluble inorganic phosphorus (SIP), (2) soluble organic phosphorus (SOP), (3) particulate inorganic phosphorus (PIP), and (4) particulate organic phosphorus (POP). The mass balance starts in an abiotic condition, with PIP dominating. However, in the presence of bacteriological activity, the dominance shifts as the PIP hydrolyzes to SIP. It is probable that the SIP (e.g., phosphate) enters cells primarily as SOP that is then polymerized into long chained polyphosphates as PIP. The sequence of phosphorus movement is therefore PIP to SIP to SOP to POP and then to PIP again. In a mature consortial biomass, any surplus phosphorus is cloistered as PIP. The ratio of these four phosphorus forms may be used to determine the state of maturity of a consormial biomass as inorganic forms of phosphorus become sequestered within cells as polyphosphates.

Nitrogen is a very dynamic nutrient essential in the production of proteins and nucleic acids. There is a very powerful mass balance created between the biosphere and the atmosphere in which similar amounts of nitrogen are fixed from (nitrogen fixation) and released to (denitrification) the atmosphere. Denitrification is primarily a broad spectrum reductive bacteriological event while nitrogen fixation is a combination of natural microbiological, geochemical and electrolytic activities.

Nutrient chemical analysis of a potentially bacteriological active liquid or solid sample has little value unless it includes determinations of total phosphorus, organic carbon, and nitrogen. Fractions of those nutrients are likely to be affected by the bacterial activities within the material investigated. Sequential sampling with analysis of these fractions may be able to project the activity levels.

Inorganic chemical analysis may relate to the bacteriological activity in the material under test but may relate more to the ORP of the sampled material and the conditions under which bioaccumulation of these chemicals occurs. Iron is possibly the most diagnostic inorganic based on bioaccumulation of ferric forms on the oxidative side of the ORP front. Geochemistry performed on a series of samples taken along a known ORP gradient from oxidative to reductive may reveal that different elements are concentrated at different points along the gradient. For example, at a severely bio-fouled well in Armstrong, Ontario, there was a successful attempt to use a blended chemical heat treatment (BCHT™ produced by ARCC Inc. of Port Orange, FL) to shock, disrupt, and disperse the biomass around the well that caused losses in water production. A sequential release of inorganic elements occurred during the block surging treatment. The order of release (earliest from the most oxidative progressing to most reductive conditions where bioaccumulation occurred) was Fe – Cu – Al –Ti – Mn –Cr – Mo – Cd – Ni – Co – Ba – V –Pb. These were the only elements analyzed routinely in water containing the dispersed biomass released by surging following treatment. These observations indicate that consortial biomass extending across an oxidative–reductive interface may form a natural "chromatograph" as different elements are bioaccumulated at their optimal ORP value ranges.

The significant implications from this potential to form a natural chromatograph are driven by the different bacterial communities sequentially accumulating these inorganic elements. Over time, a biomass will reach a point at which it can no longer maintain a dynamic integrity with its surrounding environment and it will experience partial collapses. When these collapses occur in a water well, the released bioaccumulates will appear in the produced water. Iron bioaccumulated in the natural filters around a well, because of its position close to oxidation, will be one of the first inorganics released into the water. As the more reductive zones along the biomass begin to reach saturation, they also will collapse, releasing other elements into the water. The sudden failures of old wells after years of service may be directly attributable to failures of bacterial communities within the biomass to continue to retain these bioaccumulates.

5.3.5 Water Form and Content

Little attention is generally paid to the form of water beyond defining the activities in liquid water, but water form blurs the borders between water functions (viscosities) inside and outside cells, primarily because of the production of extracellular polymeric substances (EPS) by bacterial cells that allows water to be held outside cells. Water is most bacteriologically active in the liquid phase; it is commonly present in the environment as a solid (ice) or gas (steam).

In reality, bacteriological protective functions commonly involve liquid water that provides a virtual barrier, particularly when bonded by EPS. Generally, EPS structures impact on the viscosity of the water within the structures and also through sheering effects that create biocolloidal organic materials dispersed in the water. Water viscosity measured in log cP is used in the grid atlas to define the major influence of this water characteristic on bacterial consortia.

5.3.6 Odors Emanating from Consorms

Different consorms are known to produce distinctive odors that indicate particular groups of bacteria. Odors can aid in the initial determination of the types of dominant bacteria in water. To detect these odors, the water must be collected in an odor-free container (e.g., a sterile and glass-sampling jar). The water should occupy roughly 50% of the total volume of the container and the container should be sealed. Vigorous shaking will create an aerosol of water droplets that should harbor some of the odorous material. Clear the nasal passages with two or three deep breaths, loosen the seal (unscrew and lift the cap) on the container, and gently inhale. Any odoriferous chemicals are likely to be in the gaseous phase and hence more detectable. Repeat to confirm the type of odor. If no odor is detected, warm the sample to roughly 45°C and repeat the smell test. The higher the temperature, the greater potential for the more volatile odors to be detected.

Certain odors may be linked to the activities of different groups of bacteria. This characteristic may be very useful in reaching an early decision as to which bacteria groups may cause the problem and should be further investigated. The major odors generated by microbial activities in water are listed below.

Rotten egg odor is commonly generated by anaerobic bacteria functioning in oxygen-free environments and reducing sulfates or sulfur to hydrogen sulfide or breaking down sulfur-containing amino acids in proteins to hydrogen sulfide via by proteolysis. Samples emitting this unpleasant odor may contain bacteria that use one or both reactions. Caution must be exercised when smelling this gas. The nose may become quickly saturated with the potentially toxic hydrogen sulfide and no longer able to detect its presence. One conclusion that could be drawn from the detection of hydrogen sulfide is that a sample bearing this gas almost certainly originated from oxygen-free (reductive) zones where organic nutrients were present. Aged well samples that rest for prolonged periods may generate strong rotten egg odors since the environment has become reductive and SRB have become dominant.

Fish smell is a very subtle odor that requires careful screening. The smell is commonly encountered around fish retail outlets or processing plants. Many pseudomonad bacteria frequently generate these off-odors during periods of intensive growth. These bacteria are aerobic (require oxygen) and utilize specific organic nutrients. Where pollution of a consorm environment involves specific compounds such as those associated with a gasoline release, aerobic bacteria may become dominant in the consorm.

Earthy smell resembles smells emanating from healthy soils, Some microorganisms belonging to the Streptomycetes (mold-like bacteria) and Cyanobacteria (algal-like bacteria) produce these earthy odors that are grouped in classes called geosmins. When the odors are detected, it is probable that an aerobic consorm involving one of these groups of bacteria is present.

Fecal (sewage) smell is generated by enteric bacteria commonly found in fecal material typically present in raw sewage and septic fecal wastes. These bacteria are known as coliforms. The presence of such odors in water is a very strong indication that fecal pollution is present and coliform testing is urgent because an acute hygienic hazard may exist.

Fresh vegetable smells may be produced by green algae, diatoms, desmids, and yeasts. Typical odors resemble fresh vegetables such as lettuces and cucumbers and indicate recent entry of algal-rich waters during bloom formation. Sour odors of rotting vegetation indicate algal blooms in advanced stage of decay. Large bacterial populations are expected in such waters.

Chemical smells such as those from gasoline and solvents are likely to originate from pollution rather than from the groundwater system where they may have been generated by microbial activity. If such odors are observed, a detailed chemical analyses for BTEX (benzene, toluene, ethylbenzene, and xylene) and/or hydrocarbons should be a very high priority to minimize health risk. If necessary, a gas mixture composition can be determined by a gas chromatography–mass spectrometry (GC-MS) analysis of headspace sample.

5.3.7 IDENTIFICATION OF RECOVERED BACTERIAL CONSORMS

In microbiology, a common Achilles heel is the manner in which a sample is taken to recover elements of a bacterial consorm. In many sciences, sampling methodology receives minimal attention. In the water industry, one issue is whether turning on and sampling from a gate valve provides a valid water sample. If the sampled system is biofouled to any extent, it is compromised by the microbial consorms in the region through which the water was drawn for sampling. Traditionally the reductionist approach to microbiology involves investigation at the cellular and molecular levels. Such an approach automatically involves the isolation of specific culturable cells followed by identification of function and origin based on molecular techniques.

The real challenges concern selection of a sampling site and ensuring adequate representation of the microorganisms in the consorm. If the sampling site appears reductive, it can be projected that most of the living cells in the consorm will be bacterial. If the sampling site crosses an oxidative–reductive interface going from reductive to oxidative, one can expect a broader range of bacterial communities present; capping on the oxidative side may be associated with fungi, protozoa, and higher organisms. If the interface is crossed, very different organisms will be present on the oxidative side, possibly in much higher population densities than those on the reductive side.

Choosing a sampling site within a bioconcretion is more challenging because of the cloistered positioning of consormial clusters in the biomass. Random sampling may show large differences among samples, based on whether or not the sampled intrusion extended into a "hot bed" of consormial. Good examples are rusticles that grow on deep-ocean steel shipwrecks. Between 50 and 99% of the biomass volumes in the rusticles may be virtually free of microbial activities. As a rusticle ages, the percentage of inanimate void material rises and thus the probability of recovering an active consorm would drop from one in two samples to one in a hundred sampled sites. Due diligence of sampling is required to obtain a large enough sample to represent the biomass and ensure enough duplications to find consorms dispersed in a bioconcretion.

5.3.8 TIME-SCALED MATURATION

General concepts of bacterial growth often envisage that the cells simply float or "swim" in water. In reality, bacteria tend to attach onto surfaces and then colonize the surfaces as biofilms. The bacterial cells tend to be negatively charged and are attracted to positively charged surfaces. They then become anchored to the surface by extending organic polymers (EPS, long-chained molecules) to achieve the primary attachment. Subsequently the cells reproduce and colonize the surfaces via mechanisms such as jumping and tumbling or they may simply clump to form a microcolony on the surface.

Biofilms respond to stresses during normal maturation cycles. From studies of biofouled laboratory model water wells (1 liter capacity), a sequence of events followed a basic pattern in developing iron-related bacterial biofilms. Initially, a biofilm contains a randomized mixture of a wide variety of microorganisms. Over time, however, these organisms stratify into distinct parts of the biofilm, become dominant, or are eliminated from the consortium forming the biofilm.

It is commonly considered that a biofilm forms the primary biological interface with water passing over the surface. As the water passes, chemicals are extracted and concentrated within and around the biofilm. These chemicals may be categorized as nutrients and bioaccumulates. Microbes use nutrients for growth and reproduction. Nutrient concentrations do not usually build up indefinitely. However bioaccumulates are present because they are not used and thus accumulate within the polymeric structures of the biofilm. They include nondegradable (recalcitrant) organics and various metallic ions such as iron, manganese, aluminium, copper, and zinc. The role of these bioaccumulates appears relatively passive but they may reduce the risk of predation by scavenging organisms. During maturation, a number of distinct events occur. They may be categorized into sequential phases after the formation of a confluent biofilm:

1. Rapid biofilm volume expansion into interstitial spaces with parallel losses in flow
2. Rapid decline of biofilm resistance to flow, causing facilitated flows that may exceed those recorded under pristine (non-fouled) conditions
3. Compression of biofilm volume, facilitating flow
4. Expansion of biofilm with periodic sloughing increases resistance to hydraulic flow (erratic flow); repeated cycling of harmonic fluxes created by facilitation, compression, and sloughing (sheering), ending in a stable period
5. Interconnection of neighboring biofilms, generating semipermeable biological barriers within which free interstitial water (conduits or channels) and gases (voids) may be integrated to form an impermeable barrier (plug)

During increased resistance to flow, the polymeric matrices forming the biofilm may have extended into the freely flowing water to cause radical increases in resistance to hydraulic conductivity. As resistance increases, some of the polymeric material along with incumbent bacteria will be sheared from the biofilm by the hydraulic forces. Such material now constitutes suspended particulates and forms "survival vehicles" through which the incumbent organisms move to colonize fresh econiches. In laser-driven particle counting studies, these suspended sheared particles had diameters ranging from 16 to 64 microns (in-house data). In some cases, the variability in particle size is narrow (±4 microns) related to mean particle size. Maturation of animals, plants, and also consorms involves specific stages. For consorms, the stages are acquisition of controllable space; management of water, nutrients, and gases moving though the space; intelligent integration of active microbial components to allow efficient control and management of the colonized space; and finally, structural disintegration commonly dominated by relatively useless accumulates, spent cells along with competition from neighboring organisms.

6 Bacterial Consormial Challenges

6.1 INTRODUCTION

The total mindset in microbiology is fundamentally reductionist, based on emphasis on individual bacterial species and not on complex communities. Bacterial activities have generally been at the forefront of the public view of disease. Bacterial diseases have impacted the development of all societies throughout history. Roman armies lost more men to diseases than to battle! Societies have always demanded control of disease and many common social practices can be directly linked to avoidance of disease. For example, the accepted practice of washing hands, particularly after going to the bathroom, reduces the risk of transmitting disease. Similarly, the custom of walking on the sunny side of the street arose from the observation that sunlight dries the ground; solar radiation therefore impacts certain bacteria and hence reduces risk of infection.

Disease-causing bacteria work alone to dominate the environment (plant, animal, or human) without involvement of bacterial consorms. Pathogen dominance threatens a host (animal or plant) and the various bacterial consorms that coexist in the host. It is in the interest of these natural bacterial consorms to prevent a pathogen from dominating and triggering stress, dysfunction, and even death of the host along with the resulting dispersion of the consorm. Most life forms face a prescribed death, usually preceded by a period of senescence. Consorms are not subjected to prescribed death because they function through adaptive cycles—they exist in a constant state of change.

Procedures by which bacterial consorms may be identified must consider periodic shifts in the functioning of the components in the consorm of interest. Microbiologists historically defined biological problems or events related to the activities of particular species. This approach is readily achievable for most plants and animals that have evolved into easily definable and reproducible units. This is also true of many protozoa and fungi although fungi are sometimes involved directly or indirectly in the activities and reactions of bacteriologically influenced consorms. Consorms possess unique features that involve flexibility and reflect the nature of integrated communities:

1. Consorms function as integrated intelligent collectives of different communities, primarily of bacteria.
2. Each bacterial strain in a community reacts to changes in its environment and this reactivity may affect the nature of the consorm.
3. Each consorm functions within a determinable habitat that may be adjacent to or overlap neighboring consorms.

4. Activities and reactions are dynamic, robust, and resilient and may affect the appearance, growth, and movement of the consorm.
5. Attempts to culture and biochemically define a consorm are likely to generate significant errors associable with the consorm's reactions to the associated sampling disruptions.
6. Electrical potentials and other physical and chemical factors that dominate the local environment may also influence the functionality of the consorm.

Bacteriological consorms are dynamic, flexible, and capable of morphing into different structures and compositions over time. This complicates scientific determinations of their forms and functions. Furthermore, neighboring bacterial conforms have the potential to become reactive; one may dominate or communities may simply merge. This dynamic interactive ability makes definition of a bacterial consorm more challenging unless it is based upon forms and functions and does not address the specific bacteria strains in a community. The "consorm" term is proposed to define such bacteriologically influenced communities, with the understanding that a consorm may not be biochemically definable but may be classified by the nature of an event it dominates.

In summary, a consorm involves many strains of bacteria occupying a specific definable habitat. Consorms, by the nature of their surrounding environment, undergo changes in the dominance of particular strains of bacteria over time. This shift in characteristics means that the identification of a consorm is more likely to be based on direct (form at site of activity) and indirect (odors and impacts on environment such as plugging and corrosion) evidence rather than on the traditional isolation, culture, and biochemical analysis steps. The prime focus in identifying consorms is determining the activities generated by the consorm and the reactions within the environment. Chapter 8 defines applicable biochemical analytical techniques; Chapter 9 defines culturing practices.

6.2   IDENTIFICATION OF CONSORMS

Consorms are complex bodies containing many strains of bacteria functioning in an environment that allows the entire community to flourish. Sampling a consorm is inevitably a challenge because a sample may or may not contain all the active component strains. Assuming all strains are present and active, the next decision is how to proceed with the investigation in a manner most likely to trigger the intrinsic activities and reactions of a consorm. A typical identification sequence is:

1. Determine that a consorm is present and active within the environment investigated.
2. Recognize the environmental symptoms that may relate to consorm-based activities or reactions.
3. Quantify the scale of the consorm intrusion into the definable environment.
4. Describe the nature of the consorm intrusion via cause (form) and effect (functional effect on environment).

5. Sample the consorm by an effective technique.
6. Identify significant activities and reactions relative to the sampled biomass of the consorm.

Stages in the identification process are discussed in detail below. Identification should involve prime objectives that relate to the identification of a consorm or the determination of a treatment method capable of managing the activities of the consorm usefully and economically.

6.3 DETERMINING PROBABILITY OF CONSORMIAL ACTIVITY

Consormial activity may be explained in physical or chemical terms as plugging, corrosion, rusting, tubercles, ochres, slimes, perforations, pitting, bad odors, and failing water quality. In some cases complex physical and chemical tests are used to explain the symptoms but rarely are bacterial causes considered. Consortial activities require space; a biomass may be dispersed (as in clouds) or concentrated (as in slimes). Much of a biomass may be water initially but that changes over time via accumulations. Encrustations and bioconcretions (rusticles, nodules, tubercles) may take the form of iron bioaccumulation or involve the corrosive pitting of steel, and pipe line or porous media failures due to plugging with biomass.

In some cases, evidence of consortial activity is obvious and commonly explained as physico-chemical phenomena. The symptoms listed above may point to a consormially driven bacteriological event. Findings must be confirmed by identification and quantification. Activity may be determined by adenosine triphosphate (ATP) 1 the Biological Activity Reaction Test (BART™) can define communities; Rapid Agitation Static Incubation (RASI)-MIDI testing for fatty acid methyl esters (FAME) allows biochemical identification (see Chapters 8 and 9).

A consormial biomass found in water (slime) is composed mainly of water bound to the extracellular polymeric substances (EPS). The bacteria release from the cells long stringy molecules called polymers that attract and hold water molecules. Slimes usually have water contents exceeding 90% by weight and are generally clear, white, or grey. These types of slimes grow at sites where water leaks form a vertical or horizontal surface film. As a slime matures, its texture may change from smooth and slippery to a tougher, more granular form (encrustation) as a result of the ongoing bioaccumulation of materials in the slime. Along with the encrustation, a slime may change color to white (carbonate dominance in the bioaccumulates), red or brown (iron bioaccumulation in oxidative conditions; ferric salts dominate), and grey to black (more reductive conditions; iron sulfides and/or carbonates are bioaccumulated). All these consormially influenced forms of slime will reflect different dominances of bacterial strains reflective of the environment influencing the activities. Major features of slime are color (reflecting bioaccumulated chemicals), water content (defining age of slime; water content drops with maturation), and texture (from thin viscid smooth to a more granulated form bioaccumulates are retained).

Odor may also be a major diagnostic factor for consorms observed in the environment that includes an atmospheric interface. Unlike direct observation, the identification of odors is affected by the ability of the observer to recognize them. Odors generally reflect an oxidative or reductive state of the environment. Under oxidative conditions, odors are more subtle, muted, and difficult to define. Oxidative odors are not unpleasant and range from vegetable to fruity or fishy. Fish-like odors are particularly common in consorms where pseudomonad bacteria dominate. Under reductive conditions, odors are more identifiable and often repulsive. The odor generated by hydrogen sulfide (H_2S) is the most common; it smells like rotten egg. Secondary features include blackening of the consorm from bioaccumulating iron sulfides and/or carbonates.

Other reductively generated odors include septic-like smells arising from the putrefaction of organic material and rotten vegetation and fruit odors. Reductive odors are repulsive and oxidative odors tend to be neutral or pleasant.

Consortia that emit no odor or a mildly pleasant or tolerable odor may have originated from oxidative conditions. Growths under reductive conditions away from oxygen are the most likely to be repulsive. Under some circumstances both types of slimes may encountered in neighboring environments where conditions cross from oxidative to reductive (redox front).

One common feature of many consorm-forming bacteria is the accumulation of specific metallic cations in their coatings. Iron bioaccumulates in a biomass, mostly as ferric oxides, hydroxides, and carbonates. This growth concentrates at the oxidation–reduction front within porous or liquid media where iron is present in the reduced (ferrous) or elemental (steel alloy) form.

Another metal that may dominate consorms is aluminum. The metal content of a biomass will continue to rise from 7 to 20% and even to 95% (dry weight). Generally these metal-rich structures remain porous, allowing various bacterial communities to coexist. As the metal content rises, the structure transforms from a dense slime to encrustations such as nodules, tubercles, scales, rusts, and rusticles. Eventually the bacteriological content will declines as with iron-rich biomasses. The structures terminate as various grades of pig iron.

Iron bioaccumulation also occurs at static water lines in porous media (such as soils and pack materials around water wells). In a complex microbiological reaction, a series of lateral layers of biomass growth set around the static water level. In the oxidative zones above water-saturated zones, fungi dominate at the interfaces of water, air, and underpinning surfaces. The reductive–oxidative interface commonly exhibits extensive growth of bacteriologically dominated biomass where iron bioaccumulates, primarily in the ferric form.

As iron continues to accumulate, it may form an iron pan that may be impervious to water, causing water to perch (pool) above the pan. More reductive saturated regions exist beneath an iron pan where bacteriological activity is sequentially dominated by anaerobic sulfate-reducing (less reductive gradient, G) and methane-producing (more reductive, R) bacteria. These bacteria may be recognized by color (black for sulfate-reducing bacteria or SRB) and gas bubbles (for methane-producing bacteria (MPB). These bacteria tend to compete for organic fatty acids. SRB dominate under less reductive conditions and MPB dominate in more reductive conditions. If black smelly growth appears over a zone rich in gas bubbles (with or without odor), SRB dominate over a zone of MPB.

Marine and lake sediments exist on a defined oxidative–reductive interface and often are black due to extensive activities of SRB: utilizing fatty acids and generating iron sulfides. In certain conditions, ocean sediments will extend downward 100 meters or more before entering the deeper and more reductive zones where MPB activity causes copious generation of methane. This is an example of a lateral biomass that can extend hundreds of kilometers over the sea floor below the mud line.

Iron-rich growths of biomass are commonly concretious. The structural integrity of the biomass is maintained by complex (predominantly ferric) crystalline elements that exhibit considerable porosity and larger channels through which water can flow or pool. The most sophisticated forms of ferric iron-rich bioconcretions are the rusticles that grow over steel surfaces on sunken shipwrecks in deep oceans. The bacteria first locate a site on a steel surface suitable for attachment and growth. Initial attachment often occurs on damaged steel that has become embrittled (subjected to explosives and torn). This enables the bacteria inside initiating biofilms to form a solid attachment that can then be the focus of development of bioconcretious ferric-rich growths. These steel surfaces generate oxidative–reductive interfaces where anaerobic bacteria underpin the biomass next to the steel while aerobic bacteria thrive in the more oxidative conditions created by the oxygen in the marine waters.

Bioconcretions may assume distinctive forms such as nodules (dome shapes) or tubercles (wedge shapes); they function differently from other bioconcretions in that ferric-rich capping protects their softer, less dense biomasses. This biomass activity in close proximity to the steel sets up corrosive pitting. Two major communities of bacteria may be involved in pitting and lateral fracturing corrosion: SRB and acid-producing bacteria (APB). SRB primarily generate hydrogen sulphide, causing electrolytic corrosion. APB generate acidic conditions due to fermentative outputs of fatty acids that decrease pH down to 3.0 to 5.5. This drop may significantly affect the structural integrity of metal-based alloys, particularly those containing aluminum.

Growth of any biomass can form laterally, vertically, or cylindrically as a direct result of the movement of fluids (primarily water) but other flows (e.g., oils and gases) may greatly influence the direction of biomass generation. Focused growth results from many factors including temperature gradient, ORP, compressive and turbulent shifts in fluid flow rate, changes in the electrical potentials within the environment, and nutrients (which in gas and oil flows may include water as an essential nutrient). A growing biomass finally reaches a size that interferes with the flows of fluids and gases. This interference that causes plugging (loss in flow) can also impact downstream flow characteristics. A consorm generating this plugging form of biomass may incorporate a number of definable bacterial communities that function cooperatively and occupy different locations in the biomass. Examples of plugging biomass are described below.

Golf is a major recreational industry that requires "manicured" turf grass greens to ensure optimal surfaces for the movement of golf balls. This manicuring often involves moving a green from natural soil to a synthetic medium often involving sand or sand–clay mixes. These practices can lead to the formation of a lateral black plug layer that competes with the roots of the turf grass for living space, water, and nutrients. The net effects of a successful black plug layer invasion are: (1) losses in green permeability; (2) stresses in turf grass, causing yellowing and dieback; and (3) total loss of drainage through the green, producing large patches of denuded turf.

Black plugging, like iron pans, exemplifies lateral plugging that is relatively easy to recognize. Black plug growth occurs just below the surface of a green, generally at depths of 10 to 150 mm.

Water well plugging may constitute a single locational biomass forming around the entry points of the water into the well (at slots of perforations in the casing) or as a series of cylindrical columns of consortial biomass created by bacterial communities growing around the well. This type of plugging results in reduced flows through the impacted regions that can lead to degrading water quality because of the impact of the biofouling biomass. Water well plugging may be the result of removing from or adding chemicals to the groundwater flowing through the zone and/or a diversion that changes the direction of flows entering the well. In this case, the water demand may move to flowing water from a different source that may have very different characteristics.

Biomass growth locations depend on the flow patterns and chemistry of the water; major factors include the oxidative–reductive interface, amounts of compressive flows, nutrient loading, and pH. Temperature does not appear to exert a major effect on plugging because a bacterial consorm causing plugging would have adapted (over time) to the specific ambient temperatures. Biomass can also grow in crude oil, mainly as bacterial consorms that can "mine" the water carried in the oil. Plugging of oil transmission pipelines can occur when a bacterial consorm succeeds by (1) extracting sufficient water from the oil to bind it in a manner that allows bacterial activity in the water-filled matrices; and (2) generating structures that protect the consorm from disruption.

Clearly, the challenge of increasing the water content of oil is lower (e.g., 0.2%) and the fluid flows are fast. Conditions in a pipeline are likely to be very reductive, restricting bacterial activity to strictly anaerobic forms. Deliberate or ambient electromagnetic effects may generate conditions in which the electrolysis of the water bound in the consorm may lead to localized shifts in the ORP toward more oxidative values suitable for the growth of aerobic bacteria. Protection of a bacterial consorm active within flowing oil may be enhanced by the synthesis or collection of asphaltenes or ferric salts that would form black dense masses within which the bacterial consorm could survive. This black "goop" will interfere with the rate of movement of the oil through the impacted transmission pipe, leading to higher energy costs to pump the oil, reduced efficiencies, and greater probability of corrosion if goop contacts the steel walls of the pipe.

Natural gas wells have long been thought to face little risk of plugging since natural gas is primarily methane within a gas-filled extreme environment. If water is produced along with natural gas, one may conjecture that no significant biomass may be involved because the gas is dry. However, if produced water moves with the gas, bacterial consorms have the potential to be active within the gas well environment and they generate plugging of the gas conduits to the well. As a result, natural gas production would decline as gas was diverted to compensate for the plugging of the conduits. Dry gas coming from a well does not mean guarantee that no biomass is present downhole in the gas well; it could simply mean that the biomass is further away from the bore hole and is locking up all water at that site, leaving the bore hole dry. Environmental conditions within a gas well are likely to be reductive and thus favor anaerobic activities that would restrict the microbiology to bacteria able to function effectively under these conditions. In a well generating produced water along with natural gas, the potential is greater for a bacterially influenced biomass to create plugging that would affect the sustainability of the well.

Water pipeline failures due to plugging of pipes or porous media with biomass are common problems. Corrosion affects the integrity of steel and concrete pipes. Among all types of water lines, sprinkler systems possibly pose the greatest challenge—and are the least recognized. Fire protection in modern buildings centers on the ability of heat-sensitive sprinklers to activate, release water, and douse flames with a minimum of damage. These systems require complex networks of narrow-gauge steel pipes to carry water effectively on demand to the site of a fire. These steel lines are usually full of water that is static or being recycled with makeup water needed to meet any covert leakages in the system. The risk is that bacterial consorms in these systems may grow, impose plugging risks, and cause pitting corrosion at specific sites. The danger of plugging is that the demand for water created by triggering the sprinkler system would not be met because the upstream plugging biomass would slow or stop water flow. In both cases, the sprinklers would be useless for fighting fires.

Corrosion is essentially a daughter product of a growing biomass attached to pipe walls. It creates conditions that weaken and eventually perforate the walls. Corrosion may result from pitting (linked to the bacteriological generation of hydrogen sulfide as the instigator of electrolytic corrosion) or a more generalized structural failure arising from localized acidic conditions caused by APB. Low or slow flows coupled with stagnant or slow water movement can create conditions favorable for bacterial activities. Another important feature is the electromagnetic radiation emanating from transformers. It can create charges along water pipes and thus generate focused sites for bacterial activity.

Bacterial consorms can exist in natural porous media such as soils and interact with plant roots. This activity may lead to a barter-and-exchange system between the roots and microflora on a scale that may exhibit enhanced microbial activity through several diameters of root zones—a zone of influence is known as the rhizosphere (see 2, 16-21, RZZ). The zones of bartering and exchange contain a complex weave of microbial communities. Major agents for barter include nutrients and water (from soil) and oxygen and surplus chemicals (from plants via roots).

Location is the key factor in determining the nature of bacterial consorms and their component communities. One key location to the generation of biomass is the oxidation–reduction interface (redox front) measured in millivolts (mv). A positive reading indicates aerobic (oxidative) conditions and a negative reading suggests that anaerobic (reductive) conditions dominate.

6.4 SYMPTOMS OF CONSORMIAL INTRUSIONS

Symptoms associated with bacterial consormial intrusion into an impacted environment are influenced by what would be tolerated as normal and what would be perceived as a threat to that environment. Threat is a double-edged sword: it could be considered from the exploitive nature of humans or considered a potential threat to the natural functioning of the environment. Economics rules the former and conservation governs the latter. Conflicts can therefore be clearly seen as significant events.

Perhaps a major failing in so many engineered designs in natural environments is the lack of baseline bacteriological monitoring and perhaps a maintenance manual. Baseline data indicates the development of significant symptoms that impact the data. For example, a new water well is ideally positioned to generate a baseline for production (specific capacity). All subsequent production assessments can use that specific capacity value to determine whether plugging has impacted the well. Risk analysis involves using baseline data to determine the onset of significant symptoms. A baseline should be comprehensive and include bacteriology data to validate checks. Most symptoms are recognized only when they become obvious (e.g., leaking of corroded pipe leaking, flaring gas; inadequate well flows; dieback of turf grass). By the time symptoms are obvious, the damage already may be so serious that the costs of rehabilitation will be very high (if achievable).

Little attention is paid to bacterial consorms that function within the bodies of animals and plants. In general plants are not considered to carry any bacterial loading. In animals they function only at focal sites such as the intestine where consorms participate in the digestion and breakdown of foods, reprocessing of waste materials (created directly and indirectly by the consorms), and the expulsion of waste solids. The gastroenteric tract is the site of most bacteriological activity involved in digestive functions. Other roles of consorms in plants and animals have yet to be evaluated.

6.5 QUANTIFICATION OF CONSORMIAL INTRUSIONS INTO ENVIRONMENT

Consorms are dynamic and react to changes in environmental conditions. They cannot be identified as specific (homeostatic, single steady state) strains. More functional activities and reactions may be generated using homeorhetic systems (using more than two steady states in the consormial life span). In nature, a biomass dominated by bacterial consorms may not necessarily maintain a constant ratio of integrated strains. Indeed, the total number of strains involved in a homeorhetic state may be as few as 8 in a highly eutrophic (nutrient-rich or selectively polluted state) or as many as 60 or more strains under oligotrophic environmental conditions where nutrient levels are low and may create stressful competition but no group of strains would be able to easily dominate (as in eutrophic conditions).

The nature of a bacterial consorm may be defined by its location, form, and function—and not necessarily by the nature of its individual bacterial strains. Location is defined by the form of attachment to a solid surface that may provide an observable and quantifiable biomass. If the solid surfaces are integrated into a porous medium, such direct observations would be impossible and precise characterization more challenging. Quantification would be indirect by determining the impact of the biomass now on the movements of fluids (groundwater or oil) or natural or atmospheric gases. In such cases, the rate of fluid movement would decline under constant conditions a biomass effectively reduced or blocked flow. This event constitutes a type of plugging. If a consormial biomass is attached to a solid surface, direct quantification may be achieved by direct size measurements (average length, width, and depth). An attached consorm may also be identified by its color, texture, and form.

The color of a bacterial consorm is diagnostically important because it may relate to environmental conditions. For example, ORP and chemistry provide early indications of a consorm's habitat. Black indicates high levels of sulfides and carbonates with iron being the most common cause along with hydrogen sulfide under reductive conditions. Reds and browns are dominated by ferric forms of iron under oxidative conditions. White or clear habitat indicates that conditions are probably oxidative but with little iron content that produces red and brown hues. Yellow is usually the product of dominant bacteria that naturally produce such pigments, although the color may result from bioaccumulation of nondegradable yellow pigments. Green is associated with plant chlorophyll and this type of consorm is likely to occur in daylight under oxidative conditions and potentially include cyanobacteria (blue-green algae) and/or microalgae. Purple and red pigments can appear where bacteria are dominated by strains that produce these colors under reductive low light conditions (as purple sulfur bacteria) or under oxidative conditions in which various alpha two bacteria dominate by producing violacein (violet) or prodigiosin (red). These colors may change, particularly when samples are taken and different environmental conditions apply.

Texture is another important diagnostic tool related primarily to the water content and porosity of a consorm. Two common forms of attached bacterial consorms (nodules and tubercles) exhibit low-porosity, high iron content outer shells. The shells provide a relatively smooth, usually brown, outer surface. Nodule growth is dome-shaped. Tubercle growth tends to be ellipsoid with extensive fracturing at high points. The fractures are commonly filled with exudations from the growths within the biomasses inside the tubercles.

Other iron-rich consorms that have relatively impermeable outer walls are the rusticles that grow on steel plates on sunken shipwrecks. The outer walls are often layered into three to five strata, often have several fungal components, and are permeated with ducts that extend into the inner biomass. While nodules and tubercles contain relatively amorphous gelatinous biomasses, rusticles have complex water conduits leading through ducts to centralized water channels within.

A fourth type of iron-rich consormial growth generates lateral interwoven layers of low porosity walls that provide a flaky appearance (encrustation). Instead of a centralized biomass, bacteria live on the undersides of the flakes. Manganese nodules, found on the ocean floor, exhibit this layering of flakes that form into spheres, but much of the bacterial activity occurs on the undersides of the flakes.

Porosity is a measure of the water-holding capacity of a consorm—the innate holding capacity of water within the consorm or the ability of water to pass through the growth. While the innate holding capacity may involve only slow movement of water through the growth, water may move relatively unimpeded through flaky types of growth.

6.6   CAUSES AND EFFECTS OF CONSORM INTRUSIONS IN IMPACTED ENVIRONMENT

The causes of consormial intrusions are dictated by local environment conditions that should be considered during the identification of the bacterial communities in a large integrated consormial biomass. Effects of consormial intrusions are only recognized when the impacted environment displays symptoms of dysfunctionality. If no effects are registered, the intrusion will remain unrecognized. This limits the scope to identifying only biointrusions that exert observable or significant effects on the environment. Humans do not consider a problem until a dysfunction is readily apparent. When bacteria play normal roles within what is perceived as a "normal" environment, we have no reason to identify them beyond gathering baseline data.

Focusing of a bacterial consorm in a given environment usually begins by occupying a position along a reductive–oxidative interface. Focusing determines sites along a set of constantly changing gradients where particular consorms or bacterial communities become active. The optimal site along the gradient for one consorm may not be the same for others. If a site furnishes optimal conditions for more than one consorm, competition will develop and may result in assimilation of elements of both consorms into a single and unique (unified) consorm, or one consorm may dominate, causing destruction or assimilation of weaker consorms. Competition may allow faster utilization of nutrients, oxygen, and suitable attachment sites. Antagonistic relations among consorms may readjust dominance.

Several environmental factors may affect the nature of the bacterial consorms active within a given environment. One major requirement of exploratory activities is to ensure that exploration is conducted within a temperature range at which a consorm was observed to be active. Moving samples from the sampling site to the site of investigation can severely compromise the ability to determine consormial activities with any precision. Cultural investigations require optimal temperatures. Common optima found in liquid water-dominated environments are 4, 12, 28, 37, 58, and 82°C. When investigations proceed within a narrow range, conditions close to the natural environment temperature (generally within ±2°C of optimum) will trigger effective activity.

For the optimal temperatures listed above, a range beyond ±2°C may be too narrow and impose major barriers to consormial activity occurring between 14 to 16, 30 to 32, 42 to 45, 60 to 65, and 70 to 75°C. While these are general barriers applying to different bacterial consormial groups, there are always exceptions. For example, the temperature of floating pack ice in the Arctic Ocean may range from –30 to –4°C and seems to suggest that 4°C may be considered a high temperature for examining ice-bound consorms. Many bacterial consorms found in ice-bound sites have optimal temperatures in the 4 ± 2°C range. Bioconcretions can form solid water (ice) at temperatures as high as 7°C so the presence of ice does not preclude bacterial activities.

Incubation times before the recognition of activity for these different temperatures relate to the activity of the components in the consorm. Incubation time does not relate to temperature; usually the higher the temperature, the shorter the incubation time. It is possible to see extensive activity in 24 hours at +4°C on an ice sample continually impacted by an upstream sanitary waste water discharge. Also, bacteria cells often have exterior polymeric boundary layers that act as buffers to prevent physical impacts from pressure changes.

Atmospheric pressures can also be considered to affect activities of consorms, plants, and animals. Vulnerability of plants and animals to pressure is closely related to the complex nature of cell structures within tissues. Cells may be vulnerable to pressure shifts that may disrupt their integrity. Bacterial cells, unlike those of eukaryotic plants and animals, contain few structures and are far less vulnerable to pressure change; also bacterial cells walls are simple and relatively unstructured. The structures within bacterial consorms are basically associated with accumulates within the EPS outside cells; very few structures in cells are vulnerable to pressure changes.

The flexibility and adaptability of bacteria individually and collectively to pressure changes indicate that pressures are not as significant as other factors. Essentially all the alpha bacterial groups are positioned to function at specific pressure regimes; the dominant factor affecting growth is not pressure; it may be temperature, water distribution, ORP, availability of nutrients, or biocolloidal material. Because of these other factors, alpha group bacteria are not primarily controlled by shifts in baristatic or hydrostatic pressure although their positioning in the biosphere would certainly provoke that possibility.

Location within the environment dictates to an extent which alpha group bacterial consorms are active. For example, alpha one bacteria tend to dominate in low atmospheric pressures; alpha two types function more effectively at normal atmospheric pressures. Alpha three cation bioaccumulators and concreters tend to function at the reductive–oxidative interface that transitions conditions from oxidative (above) to reductive (below). At the interface, hydrostatic pressures begin to rise. Generally these environments involve combinations of gases within fractured or porous media. Alpha four bacteria are set more deeply as lateral layers in the reductive zones of sediments and rock structures associated with the deeper regions of the crust. The very significant increases in hydrostatic pressures would be lethal to all plants and animals but alpha four bacteria can survive, remain active, and generate hydrocarbons. Alpha five bacteria also form at the oxidative–reductive interface, more on the oxidative side. The degradation function primarily uses organics along with petroleum hydrocarbons, volatiles, and gases that may be moving back to the surface from the deeper regions of the crust where alpha four consorms had been active. The activities of alpha five bacterial consorms—degrading upwardly mobile hydrocarbons—may constitute the biggest natural bioremediation project undertaken by Nature!

Alpha six bacteria are speculated to be active under extremely high hydrostatic pressures in the hot transitional phase where water recycles between liquid and gaseous phases. In many ways alpha six bacteria resemble alpha one types in that their activities occur within bacteriologically nucleated water droplets. In our natural preoccupation with our bacterial neighbors in the surface biosphere, we focus the most attention on the alpha two groups. These bacteria can be sterilized (killed) by elevating the temperature to 121°C at a pressure of 103 kPa (kilopascals). While these conditions may be appropriate for sterilizing most alpha two and three bacteria, they may not be appropriate for controlling the speculated alpha six consorms.

ORP constitutes a major influence on the activities of various bacterial consorms. This is because the interface from a positive (oxidative) condition to a negative (reductive) condition (both measured in millivolts) is a critical boundary between respiratory and fermentative activities, respectively. This redox front may also be considered a transitory zone between oxygen-dominated (above) and hydrogen-dominated (below) environments and serves as a fundamental division of bacteria into two major groups in which the alpha one and two groups are essentially on the oxidative side (–50 to +400 mv) while alpha threes are often at the interface (–50 to +200 mv) and alpha fours and sixes range between –50 and –800 mv. A major exception is the alpha two group active in oxygen suppressed environments (e.g., in gastroenteric tracts and muds). They thus tend to function fermentatively under mildly reductive conditions.

Nutrient acquisition is a major challenge for bacteria, whether they function in respirational or fermentative modes. The function of any nutrient is two-fold: (1) creating energy through its degradation or molecular manipulation; and (2) exploitation as a molecular source for components essential to the survival or growth of bacterial cells.

Where a nutrient is exploited by bacteria as a means of creating energy through its degradation or molecular manipulation, the energy is created by the fractionation of the molecule and subsequent release of energy. Under oxidative conditions where respiratory functions dominate, catabolism of the organic molecule can be complete and generate a major daughter product, carbon dioxide. Under reductive conditions, the energy generally is released with less efficiency and the major daughter products are often dominated by fatty acids. If the feedstock nutrients are fatty acids, the daughter products may be dominated by hydrocarbons. For example, under very reductive conditions (fewer than –150 mv), alpha four consorms will dominate and generate biogenic methane, petroleum hydrocarbons, coal, and tar sands. Under less reductive conditions and in the presence of sulfates, competition would arise between the SRB and the alpha four consorms for the available fatty acids. At ORP values more reductive than –150, alpha four consorms will dominate.

Of the nutrients present naturally in environments, organic carbon forms and phosphorus are often the primary factors influencing the levels of bacteriological activity. An almost infinite variety of organic carbons can affect bacteria that would remain active and dominate in a consorm. Generally the concentration of organic carbon that can trigger significant bacterial activity can be measured in the low parts per million (ppm) range. Phosphorus is available in a limited range of forms: soluble inorganic (SIP), soluble organic (SOP), particulate inorganic (PIP), and particulate organic (POP). Unavailable forms tend cluster in the PIP class while the most active forms are tied up in the POP class. The value of phosphorus for many bacteria is shown in the high percentage of polyphosphates bound into biomasses, mostly within cells. Organic carbon levels below 10 ppm and phosphorus below 0.1 ppm can restrict the amount of bacteriological activity observable under such oligotrophic conditions.

Nitrogen is another macronutrient essential for bacteriological activities but it differs from organic carbon and phosphorus in that a very dynamic cycle moves nitrogen into the atmosphere (denitrification) and from the atmosphere (nitrogen fixation). Many bacterial groups are integrated into the denitrification function that occurs under reductive conditions. Under oxidative conditions there are two major activities: the direct fixation of dinitrogen gas under both reductive and oxidative conditions and the oxidation of ammonium generated under reductive conditions to nitrate (nitrification). The alpha two consorms include specialized bacteria that can cause nitrification or nitrogen fixation; many bacteria can cause denitrification.

In relative importance, carbon (mostly organic forms) dominates nitrogen which is required by cells to a greater extent than phosphorus. A typical carbon:nitrogen:phosphorus (C:N:P) ratio is 100:10:1. For optimal growth and activities, a bacterial biomass should possess similar ratios. If a distortion in favour of phosphorus occurs, the additional phosphorus is most likely to be in PIP and POP forms as polyphosphates. The natures of the total carbon, nitrogen and phosphorus nutrients in an environment can reveal which bacteria consorms will dominate.

pH is another major factor for determining types of bacterial consorms that would form actively. It shows whether an environment is acidic, neutral or alkaline. Bacteriological activities have been observed over the full pH spectrum from 0 (extremely acidic) to 14 (extremely alkaline). Many bacteria function over wide pH ranges due to the buffering capabilities in a biomass created by EPS (extracellular polymeric substances). Most consorms function within a pH range of 6.5 to 8.5, with optimal growth at 7.5 to 8.0. Many bacterial consorms active at higher (alkaline) pH values tend to be alpha threes that generate iron- or carbonate-rich concretions to provide some protection from high pH values. Reductive conditions tend to create fermentative activities whose acidic daughter products are fatty acids. Reductive conditions can lower the pH into the 3.5 to 5.5 range. Under oxidative conditions, oxidation particularly of sulfur and sulfides to sulfates and/or sulfuric acid may occur. The terminal pH values can be as low as 0.5, 1.5 or 2.0. Generally the terminal pH value indicates the types of bacteria involved in a given oxidative event. For example, terminal pH values between 0.5 and 2.0 indicate oxidative conditions with a potential for dominance by sulfur-oxidizing bacteria including species of Thiobacillus, also known as Acidothiobacillus.

6.7   CONSORM SAMPLING PROTOCOLS

Sampling an environment to recover a functioning consormial biomass presents significant challenges originating from the need to disturb the environment to obtain samples. Clearly any attempt to determine the nature of a consorm must minimally involve an impact on the environment containing the consorm and maximally cause total disruption of the consorm-dominated biomass. Any sampling protocol must therefore accept the occurrence of such disruptions and the inevitable bias of the resultant information. If a sample must be taken from a porous or fractured medium, coring may be a possible approach. If coring is not possible, the sample should be limited to using water passing through the zone of interest as the carrier of the bacterial biomass. The water will reflect only bacteria that must be (1) released into suspensions within transient flowing water; (2) dispersed enough to be enumerated separately; and (3) able to be detected using the selected analytical system.

Sample bacteria so recovered will have to adapt to temperature shifts that may occur during the move from the site to the sampler, within the sampler, and then in an analytical forensic microbiology laboratory. These temperature shifts may cause trauma to some of the bacteria within the consorm; this will bias the data toward bacteria that survived sampling and analysis. The convenient and accepted use of agar spread plates often incubated under highly oxidative conditions at higher temperatures than the sampled site (e.g., 28°C for a sample site functioning at 9°C) would further distort the real value of the data gathered. Agar spread plates are known to be very selective due to the traditional reliance on distinct colonial growths as primary features in the detection of bacterial types. This further restricts the sensitivity of test methods. Because a sample contains only bacteria that naturally occur in the water and those released from the biomass into the water, the value of the data becomes questionable. Similar reproduction of the data through repeated results merely indicates that the protocols employed may contain a common set of built-in errors and may not reflect science although they would validate precision.

Clearly, a major concern for achieving confidence in data developed for a sample is its relevance to the active consorms at the originally sampled site. This may be corrected to an extent by the direct analysis of the bacterial consortium at the site or as quickly as possible. Technologies now in development (see Chapters 8 and 9) allow this to be achieved primarily by direct assessment of the high energy storage capacity of a consortium by testing for adenosine triphosphate (ATP) or using fatty acid methyl ester (FAME) analysis to detect the activity levels of samples. DNA analysis can detect the principal components of a consormial community in a sample.

7 Detailed Identification of Bacterial Consorms

7.1 INTRODUCTION

Bacterial consorms are not defined by sets of microbiological and biochemical protocols because each one is uniquely bounded by the environment within which it forms and performs a significant task. The first level of identifying a consorm is therefore to evaluate conditions where it is active (grows). This provides information on the environmental interface where growth occurs. The information gathered provides enough data to define the type of consorm at the first level (alpha type and most probable definition).

Essential data include the average oxidative–reductive potential (ORP) within the environment that would define the biomass as primarily oxidative (aerobic), transitional—between oxidative and reductive (facultatively anaerobic), or reductive (anaerobic). The next stage is to determine the abundance and viscosity of the water in the environment. Water may be free flowing and clear; turbid to more viscid based on biocolloidal content; waters that are basically entrained in the biomass and show little evidence of films; and highly viscid waters that appear only within certain biological (e.g., cell) or chemical (e.g., clay or mud) entities. Water may also simply be dispersed within a gaseous phase such as air or steam. Such dispersed liquid water particles may still be capable of supporting bacterial activities!

Traditionally temperature was considered a major factor in selection and growth of organisms. For most bacterial consorms, environmental temperature is constant unless directly or indirectly affected by diurnal or seasonal rhythms. In the latter case, the rhythms are experienced mainly in the surface biosphere and where organisms show greater tolerances to temperature shifts.

Surface biospheric bacteria exhibit some ability to accept broader ranges of temperature through which activities and growth can be achieved. This can be reflected in the temperature range around the optimal temperature over which bacteria will be active. Surface biosphere bacteria must adapt to diurnal and seasonal temperature shifts. Consequently, the activity temperature range around the optimum would be broad (ranging from 10°C or 5°C either side of the range). Bacterial consorms that are not normally impacted by diurnal and/or seasonal temperature changes (such as deep biosphere bacteria) are likely to have reduced ability to adapt to these shifts. Little research has been conducted on such bacteria but they are thought to experience a narrower temperature range of activity (e.g., ranging from 5°C below to 3°C above the optimum). Table 7.1 lists the different consorms by focal point locator. Table 7.2 lists them alphabetically by their three-letter codes.

TABLE 7.1
Definition of Bacterial Consorms by Focal Point Locator on Bacterial Positioning Grid

1, 17-15	CMT (Comets)*	2, 24-13	TCT (Thatch)
1, 22-03	CLD (Clouds)	2, 15-13	TBD (Turbidity)
1, 19-06	FOM (Foam)	2, 07-20	ULC (Ulcers)
1, 16-12	ICE (Ice)	2, 12-07	URN (Urinary tract infections)
1, 16-03	LNG (Lightning)*	2, 12-13	WLT (Wilt)
1, 19-12	SNO (Snow)	3, 10-21	BPL (Black plug layer)
2, 12-17	ABS (Abscesses)	3, 10-27	BBR (Blueberries)*
2, 13-10	BOL (Boils)	3, 03-16	BWR (Black water)
2, 19-07	BCH (Bronchitis)	3, 19-03	CRB (Carbonates)
2, 08-20	CBC (Carbuncles)	3, 05-27	CGG (Clogging)
2, 11-14	CCD (Carietic condition)	3, 06-16	CRS (Corrosion)
2, 15-07	CLS (Cloudiness)	3, 18-25	CCR (Concretion)
2, 11-18	CHR (Cholera)	3, 15-21	ECR (Encrustation)
2, 16-13	CLB (Coliform bacteria)	3, 20-28	FRD (Ferric-rich deposits)
2, 15-18	CLW (Colloidal water)	3, 22-18	IPN (Iron pan)
2, 18-13	DRR (Diarrhea)	3, 18-14	LSL (Lateral slime layer)
2, 10-12	FCB (Fecal coliform bacteria)	3, 15-28	MGN (Manganese nodules)
2, 07-23	FEC (Feces)	3, 18-19	OCR (Ochres)
2, 10-08	KDS (Kidney stones)	3, 10-30	PGI (Pig iron)
2, 22-12	LMN (Luminescent bacteria)	3, 03-19	PFR (Perforation)
2, 09-15	MIC (Microbiologically influenced corrosion)	3, 06-24	PTG (Pitting)
2, 18-16	NCR (Necrosis)	3, 19-21	PLG (Plugging)
2, 22-21	PMP (Pimples)	3, 19-26	RST (Rusticles)
2, 17-15	PLG (Plugging)	3, 10-10	SBL (Sand boils)
2, 23-14	PEY (Pink eye)	3, 15-17	TCL (Tubercles)
2, 18-18	PUS (Pus)	4, 15-25	BAP (Black asphaltene-rich plugs)
2, 06-18	ROE (Rotten eggs)	4, 04-20	CDR (Carbon dioxide reducers)
2, 14-15	ROT (Rotting)	4, 05-17	COL (Coal)
2, 16-21	RZZ (Rhizosphere)	4, 06-22	BNG (Biogenic natural gas)
2, 22-07	SHN (Sheen)	4, 03-13	BSR (Black smokers)
2, 23-21	SKN (Skin)	4, 01-28	OIL (Oil)
2, 11-16	SLT (Silt)	4, 07-19	PET (Peat)
2, 21-22	SLM (Slimes)	5, 15-10	GHY (Gas hydrates)
		6, 01-03	HBC (Hyperbaric bionucleating)*

Asterisk (*) = speculative.

7.2 DEFINING BACTERIAL CONSORMS BY FORM, FUNCTION, AND HABITAT

All bacterial consorms present a definable form within a habitat where they are recognized by their perceived functionality. Form indicates the shape, arrangement of parts, exterior aspects unless the form is sufficiently diffuse, and the mode in which the consorm exists. Function reflects the mode of action or activity through which

TABLE 7.2
Definition of Bacterial Consorms by Three-Letter Code and Focal Point Locator on Bacterial Positioning Grid

ABS (Abscesses) 2, 12-17
BAP (Black asphaltene-rich plugs) 4, 15-25
BBR (Blueberries)* 3, 10-27
BCH (Bronchitis) 2, 19-07
BNG (Biogenic natural gas) 4, 06-22
BOL (Boils) 2, 13-10
BPL (Black plug layer) 3, 10-21
BSR (Black smokers) 4, 03-13
BWR (Black water) 3, 03-16
CBC (Carbuncles) 2, 08-20
CCD (Carietic condition) 2, 11-14
CCR (Concretion) 3, 18-25
CDR (Carbon dioxide reducers) 4, 04-20
CGG (Clogging) 3, 05-27
CHR (Cholera) 2, 11-18
CLB (Coliform bacteria) 2, 16-13
CLD (Clouds) 1, 22-03
CLS (Cloudiness) 2, 15-07
CLW (Colloidal water) 2, 15-18
CMT (Comets)* 1, 17-15
COL (Coal) 4, 05-17
CRB (Carbonates) 3, 19-03
CRS (Corrosion) 3, 06-16
DRR (Diarrhea) 2, 18-13
ECR (Encrustation) 3, 15-21
FCB (Fecal coliform bacteria) 2, 10-12
FEC (Feces) 2, 07-23
FOM (Foam) 1, 19-06
FRD (Ferric-rich deposits) 3 -20-28
GHY (Gas hydrates) 5, 15-10
HBC (Hyperbaric bionucleating)* 6, 01-03
ICE (Ice) 1, 16-12
IPN (Iron pan) 3, 22-18
KDS (Kidney stones) 2, 10-08

LMN (Luminescent bacteria) 2, 22-12
LNG (Lightning)* 1, 16-03
LSL (Lateral slime layer) 3, 18-14
MGN (Manganese nodules) 3, 15-28
MIC (Microbiologically influenced corrosion) 2, 09-15
NCR (Necrosis) 2, 18-16
OCR (Ochres) 3, 18-19
OIL (Oil) 4, 01-28
PET (Peat) 4, 07-19
PEY (Pink eye) 2, 23-14
PFR (Perforation) 3, 03-19
PGI (Pig iron) 3, 10-30
PLG (Plugging) 2, 19-21; 3, 19-21
PMP (Pimples) 2, 22-21
PTG (Pitting) 3, 06-24
PUS (Pus) 2, 16-18
ROE (Rotten eggs) 2, 06-18
ROT (Rotting) 2, 14-15
RST (Rusticles) 3, 19-26
RZZ (Rhizosphere) 2, 16-21
SBL (Sand boils) 3, 10-10
SHN (Sheen) 2, 22-07
SKN (Skin) 2, 23-21
SLM (Slime) 2, 21-22
SLT (Silt) 2, 11-16
SNO (Snow) 1, 19-12
TBD (Turbidity) 2, 15-13
TCL (Tubercles) 3, 15-17
TCT (Thatch) 2, 24-13
ULC (Ulcers) 2, 07-20
URB (Urinary tract infections) 2, 12-07
WLT (Wilt) 2, 12-13

Asterisk (*) = speculative.

the consorm fulfills its purpose. Habitat defines the natural home of a bacterial consorm. Each consorm possesses particular features within its natural form, function, and habitat that allow it to be recognized as definably unique within the alpha system of classification.

The order presented in Chapter 3 for the primary differentiation of the alpha bacterial consorms (with more detailed aspects in Chapter 4) will define the bacterial consorms. Each consorm will be listed first by alpha group number followed by grid reference (XX-YY), three-letter reference code, brief descriptor, and finally a definition of its form, function and habitat. Speculative conforms are indicated by asterisks (*).

1, 22-03 CLD (Clouds)

Form: These are very unusual bacterial consorms in that their habitat is formed by the atmosphere and strongly influenced by meteorological events. The biomass is dispersed mainly as charged (bionucleated) water particles in the atmosphere that can be easily destabilized by changes in atmospheric conditions. The usual product of these changes is rain or snow.

Function: Clouds are essentially Nature's floating Band-Aids because they perform as active microbiological filters for bioremediation involving the removal of dispersed and volatile organics streaming up into the atmosphere from the surface.

Habitat: Clouds are present in the atmosphere in several definable forms that affect the amount of solar energy that reaches the surface. They originate from water surfaces via evaporation or natural electrolytic functions that may be accelerated by wave actions in open waters. Thermal gradients appear to play a major role in cloud formation.

1, 17-15 CMT (Comets)*

Form: Comets are large irregular objects that resemble rock concreted into ice or they may have smoother surfaces dominated by ice.

Function: For a bacterial consorm, ice can be synthesized in water-saturated conditions at temperatures up to 7°C, form a protective blanket from predation, and provide an additional site for accumulated reserves of organics and metallic cations. It is highly probable that such environmental conditions are likely to trigger bacteria to enter suspended animation states as ultra-microbacteria and prolong their states of dormancy.

Habitat: A comet passing though deep space is exposed to extreme physical conditions including low temperature vacuous states, and periodic high levels of radiation. These conditions indicate that comets contain very low or no levels of microbiological metabolic activity; comets may be biocarriers.

1, 19-06 FOM (Foam)

Form: Foam is a coalesced group of bubbles. The biofilm membranes that form the bubbles initially combine but maintain integrity of the bubbles within the group. This means foam may be multilayered. With wave action or wind, larger foam structures can be blown into the air.

Function: Foam is more likely to form in higher nutrient loading (eutrophic) conditions and can accelerate the rate of biochemical degradation within the individual bubbles forming the foam.

Habitat: Found on water surfaces under eutrophic conditions, particularly where water is blown toward the shoreline. Wave action can also create foam particularly in eutrophic waters and at the crests of waves.

1, 16-12 ICE (Ice)

Form: Ice is solid crystallized water in a frozen state. It has a marginally lower density than water and therefore tends to float. Some bacterial concretions can synthesize ice at temperatures up to 7°C. Additionally, ice consorms can release surfactants that keep some of the internal water in a liquid state and allow growth to occur at temperatures down to –15°C or lower. Such growths within ice may be recognized as discolorations or fracture points.

Function: Traditionally, ice was considered sterile and too cold to support life. However, the abilities of many bacteria to generate extracellular polymeric substances (EPS) that possess antifreeze properties allow consorms to function within saturated liquid biocolloidal voids inside solid ice. A consorm growing in ice operates within unique niches where low temperatures retard biochemical activities, at least initially until a consorm adapts to the extreme temperatures.

Habitat: Ice floats in water as solid crystallized structures. Consorms function only within small liquid-filled voids and may cause discoloration of the ice in lateral layers. Ice may also form over land masses as slowly moving or static glaciers. In both cases, consormial activities focus in the more eutrophic regions created by the admission of particulates (e.g., soil or dust) or organic leachages (e.g., sewage, landfill, or natural compost discharges).

1, 16-03 LNG (Lightning)*

Form: Lightning generally means the sudden intense and powerful discharges of electrical energy from clouds. Consorms functioning within clouds include complex communities of integrated bacteria that carry unique cationic (negative) charge potentials. Within clouds, the charged, bacteriologically dominated aerosol particles act as capacitors that may be stimulated by friction between different clouds having different consormial charges. When these biocapacitors reach a maximum charge, the charges are discharged as lightning.

Function: Essentially, lightning discharge is a terminal event when the consormial capacitors are fully charged. The function of lightning thus is to maintain the bioelectrical balances within clouds to ensure ongoing harmonized bacterial bioelectrical activities.

Habitat: Lightning forms mainly between elements of clouds that carry dissimilar cationic charges from different bacterial consorms. Lightning may travel along pathways created by these different consorms.

1, 19-12 SNO (Snow)

Form: Snow is considered unique because crystallized flakes of water displayed a different structure, generally in a fractal form radiating outward in two planes from a central axis. This meant that each snowflake only grew along those two dimensions and displayed bilateral symmetry. When snowflakes are made by super-cooling aerosols fabricated from pure cultures of bacteria (e.g., Pseudomonas species) all the resultant snowflakes have a common crystalline pattern.

Function: Because water in a snowflake is crystallized solid ice, the function of the snow relates to the way the ice was frozen rather than a mechanism that allows metabolic functioning. Snowflakes may indicate the freezing of EPS into complex and delightful crystalline forms rather than expressing active metabolism.

Habitat: Snowflakes are abundant on surfaces below freezing point. As the snow packs down, its weight creates pressure that may turn the snow into ice. Evidence to date suggests that snow forms a survival habitat where bacterial growth does not occur under normal circumstances.

2, 12-17 ABS (Abscesses)

Form: An abscess is a swollen region within a tissue that accumulates pus (a thick greenish or yellowish liquid produced by infection).

Function: Abscesses focus at infection sites within animal tissues; the microbe biomass generates sufficiently to create outward pressures on the surrounding tissues. The pressures cause shearing of the surrounding tissues that relieves the pressures by discharging part of the biomass as pus. An abscess is a localized form of infection recognized by tissue swelling that may be accompanied by elevated tissue temperature and exudation of pus.

Habitat: Localized tissue site where the infecting consorm grows in volume. The surrounding tissues provide resistance to invasion and seals the infested region (abscess). Nutrient and oxygen initially flow through the natural conduits (blood and lymph). As the walls harden in an abscess biomass, the flows of nutrients and oxygen are cut off and conditions become more reductive.

2, 13-10 BOL (Boils)

Form: A boil is an inflamed pus-filled swelling associated commonly with infection of a hair follicles.

Function: A boil is generated by invasion of the roots of hair follicles, generating an abscess-like infection, with potential for the early relief of intrinsic pressures via releases of pus within the follicles.

Habitat: Generally a localized infection in the skin associated with hair follicles.

2, 19-07 BCH (Bronchitis)

Form: Bronchitis consorms develop as localized inflammations of mucosal membranes in the bronchial tubes. Usually bronchitis takes the form of aggressive biofilms growing on the surfaces of the bronchial tubes followed by interconnections that further restrict air flows. The integrity of the infesting biomass is enhanced by chemical and/or particulate materials that form hardening bioconcretions and the bronchitis becomes chronic.

Function: To generate an infesting biomass in the bronchial tubes to create chronic or acute symptoms in a host.

Habitat: Biofilms and bioconcretions within bronchial tubes.

2 , 08-20 CBC (CARBUNCLES)

Form: These consorms cause severe skin abscesses that may develop a bright red color.

Function: Carbuncles are generated as severe infestations within skin tissues.

Habitat: Generally a localized biomass that dominates a small region of skin tissue. The host tissues and immune system cannot control the growth of the consorm but can limit its spread to the domain of the abscess.

2, 11-14 CCD (CARIETIC CONDITION)

Form: A carietic condition or decay event leads to crumbling of the infested bone. Carietic conditions frequently appear in teeth—hard, bony, enamel-coated structures that shred and masticate food. Caries compromise the enamel coating, allowing an infection to move into the dense bony tissue that forms the bulk of a tooth. After the enamel is penetrated, the dentine becomes the prime site of the infestation.

Function: To generate a slow growing biomass that gradually infests most of a tooth.

Habitat: After initial penetration of the enamel, the habitat is inside the dentine where the host's defense mechanisms have restricted access because of the density of bony tissue. Secondary infections may occur in the underlying gum and nerve tissues if the carietic condition is not controlled.

2, 15-07 CLS (CLOUDINESS)

Form: Cloudiness in this context refers to the presence of particulate matter in water that causes the water to lose clarity as the particles suspended in the water interfere with the passage of light. This can result in increased reflectivity from the particles or sorption of the light passing through the water to create turbidity. While cloudiness is commonly linked to water, the physical characteristics can be applied to the atmosphere (see 1, 22-03 CLD for details on cloud consorms). During daylight hours cloudiness is recognized by interference to light penetration arising from water-laden biocolloidal aerosol particles in the air. Both forms of cloudiness are commonly diffuse and relatively dynamic over time.

Function: Cloudiness in water is symptomatic of the upstream generation of biomass that has sheared and moved into the water phase. Water is the carrier and not the cause. The biocolloidal particles suspended in the water hold most of the living biomass. The particles remain active and generate significant adenosine triphosphate (ATP), most likely from the sheared biomass, that may be significantly supplemented with nutrients and oxygen still within the water.

Habitat: Cloudiness is a randomized collection of small particles in water that do not display stable structures. If an oxidative–reductive interface is generated within a sample, floating plates may be observed associated with the interface. Static waters may contain thread-like processes that generally connect to the headspace air interface or other surfaces.

2, 11-18 CHR (CHOLERA)

Form: Acute infestation of the small intestine that triggers severe vomiting and diarrhea; may lead to radical dehydration of the infected host that can be fatal. The symptoms arise from infection by a single aggressive species, *Vibrio cholerae*, but the role of natural consorms in supporting or controlling the infection cannot be ruled out.

Function: Domination of the small intestinal tract and radical trauma to the host, leading to vomiting, diarrhea, and dehydration that may be fatal. Transmission of the pathogen is via watery stools and vomit.

Habitat: Small intestine; infection may be triggered in a host that is nutritionally and immunologically compromised.

2, 16-13 CLB (COLIFORM BACTERIA)

Form: Coliform bacteria grow primarily in the small intestine and appear to dominate bacterial activities based on a unique evolutionary development in mammals: the young are fed high lactose content milk. This means that coliform bacteria have an inherent ability, at least initially, to live on high lactose nutrient inputs while tolerating the bile salts injected into the intestine by the host.

Function: Coliform bacteria undertake critical metabolic degradation functions in the intestinal tract that allow ready absorption of daughter products by the host. This function usually occurs without any significant deleterious impact on a host. Coliform consorms include a range of species. It is important to note that: (1) *Escherichia coli* is a common marker for potential hygiene risks; and (2) certain genera such as Salmonella and Shigella may become pathogenic to a host and cause typhoid, paratyphoid, or dysentery.

Habitat: Coliforms function well in the smaller intestine and pass from the body in feces. Many of the consorm cannot adapt to the environments and die off. Others adapt and become active in natural ecosystems (e.g., Enterobacter, Serratia, Citrobacter and Klebsiella species).

2, 15-18 CLW (COLLOIDAL WATER)

Form: Colloidal water is evenly viscid water and does not show distinctive structures when illuminated. Tipping colloidal water in a test tube reveals resistance to flow and the meniscus may become buckled away from normal.

Function: Colloidal water is likely to contain high total suspended solid (TSS) content with relatively little variability when assessed by a laser particle counter. The water is charged with EPS produced by bacterial consorm activity.

Habitat: Stagnant and slowly moving water bodies with significant total organic carbon loading that triggers the production of EPS from the bacterial biomass.

2, 18-13 DRR (Diarrhea)

Form: Excessively frequent and loose bowel movements often leading to watery stools.

Function: Dysfunctional intestinal tract characterized by destabilization of bacterial flora.

Habitat: Gastroenteric tract with focus on dysfunction in smaller or larger intestinal tract.

2, 10-12 FCB (Fecal Coliform Bacteria)

Function: These bacteria play a major role in the normal functioning of the gastroenteric tract but also appear to support some species that can become pathogenic.

Habitat: The gastroenteric tract in both healthy and dysfunctional states. The principal species (*Escherichia coli*) appears to have limited ability to adapt to natural environments. However, other fecal coliforms can become participants in natural biomasses and may register significant hygiene risks related to *Escherichia coli* when the risks are in fact lower.

Form: Fecal coliforms are grouped around *E. coli* which is known as the most suitable indicator for hygiene risk in waters and foods. A consorm is established by those enteric bacteria can be fermentatively active, producing gas at 44.5°C in the presence of bile salts and lactose.

2, 07-23 FEC (Feces)

Function: Relatively little attention is paid to the role of feces in animals that have gastroenteric tracts. Fecal material is the product of two functions: (1) dewatering to ensure that the water is retained in the body; and (2) extraction of useful nutrients from the feces before passes from the body. Fecal material has a very high bacteriological content since the voided material includes the bacterial consorms involved in dewatering and extraction.

Habitat: The colon from the cecum to the rectum where compressive forces act to dewater the stools and bacteriological activity is associated with the extraction of nutrients and to some extent the structuring of the feces.

Form: Varies with animal species and depends primarily on the need to conserve water. Birds and many insects, for example, extract so much water that the expelled fecal form is relatively hard. Animals such as fish living in an environment submerged in water do not have such extreme needs to conserve water.

2, 10-08 KDS (Kidney Stones)

Function: These spherical solid bioconcretions (stones) that grow in the kidneys have the prime function of removing nitrogenous wastes from the blood and excreting them as urine. They are essentially solid nonmetallic structures that interfere with the passage of urine from the kidneys.

Habitat: Within the kidney and passageways from the kidney to the bladder where they cause blockages.

Form: Hardened, spherical, generally lateralized bioconcretious structures that provide a protected habitat within which bacteria can remain active. Over time, the ring structures multiply as the stones expand where the consorm remains active. Spherical metallic and non-metallic structures are observed in most bioconcretions.

2, 22-12 LMN (LUMINESCENT BACTERIA)

Function: These consorms share a common biochemical pathway with fireflies. The pathway enables the direct release of energy stored primarily in adenosine triphosphate (ATP) as light by utilizing the luciferase enzyme system.

Habitat: Luminescent bacteria are most easily detected under very low light or dark conditions; they generate a pale blue light under oxidative conditions. The largest example is the deep-ocean environment. At 400 to 1,000 meters, the deep scattering layer (DSL) emits blue bursts of light when the environment is disturbed (e.g., by a submarine or remotely operated vehicle). These bacteria have also been reported active in caves and generate enough light to allow algae, mosses, and other plants to grow.

Form: Luminescent bacteria appear in biofilms or suspended biocolloidal particles under modestly eutrophic oxidative conditions. The use of intrinsic ATP to generate light has a high energy cost, so continuous bioluminescence is observed only under extremely eutrophic conditions. The DSL appears to emit light only in response to pressure changes but light may also be emitted in response to other challenges, such as risks from predators.

2, 09-15 MIC (MICROBIOLOGICALLY INFLUENCED CORROSION)

Function: Generation of biomass, usually as biofilms or bioconcretions that become destructive after attachment to surfaces that have economic value. Growth of the biomass secondarily compromises the structural integrity of the surface. Biomass activity triggers two major corrosive events: (1) generation of hydrogen sulfide that leads to electrolytic pitting and perforation in iron alloys such as steel; and (2) generation of organic (fatty) acids with a localized pH reduction that leads to lateralized corrosion, particularly in concretes. Corrosion can also be triggered by sulfuric acid generated by sulfur-oxidizing bacteria (SOB).

Habitat: Microbiologically influenced corrosion is generally seen as occurring only in reductive environments. Because biofilms and bioconcretions can grow to include oxidative zones underpinned by reductive zones, this type of corrosion always has the potential for the reductive (anaerobic) generation of hydrogen sulfide or fatty acids within a maturing growing biomass. Corrosion may also be generated under oxidative conditions in the presence of elemental sulfur or reduced forms of sulfide. The SOB traditionally considered to belong to Thiobacillus species generate sulfuric acid that reduces pH to a level that corrodes steels and concretes. High sulfur content ore tips and abandoned water-saturated mine shafts are common habitats for SOB, leading to low pH corrosive discharges.

Form: The nature of the surface corroded affects the form. For steels suffering from lateral or vertical corrosive processes, the biomass will accumulate ferric iron forms that harden outside the growth. The form of the growth in this case is lateralized, with pitting and perforation under the biomass which is itself protected by a complex ferric iron oxide and hydroxide. Coating of concrete may take the form of a patina or bioconcretion. Generally the corrosion form hardens over time due to the accumulation of disrupted debris generated by the corrosion.

2, 18-16 NCR (Necrosis)

Function: Produces death of tissue by disease or injury. Necrosis is a localized event. In disease, it may produce secondary complications for the host (e.g., tissue gangrene).

Habitat: Commonly focused in specific types of plant and animal tissue; the aggressive nature of the consorm activity leads to tissue death. The dominant event in plants is that the invading consorm has cellulolytic capabilities. In animals, a consorm aggressively produces proteolysis and possibly gases through fermentative functions.

Form: Localized in one type of tissue that may become reductive early in the infestation. In animals, the constrictive nature of gases formed fermentatively may block the supply of blood to the infested tissue, exacerbating the problems.

2, 22-21 PMP (Pimples)

Function: Small, hardened, inflamed spots that generate localized reactions (undefined infestation or irritation) in the skin.

Habitat: Localized change in the skin that shows inflammation (indicating possible infestation) and hardening (indicating changes in tissues resulting from irritation or prolonged infections with possible production of recalcitrant gases within tissue).

Form: Hardened, inflamed spots on the skin.

2, 19-21 PLG (Plugging)

Function: Physical or biological obstruction that interferes with the passage of water through porous or fractured media. Plugging is normally biological and involves a complex biomass that differentiates environmental conditions within the plugging zone (see also 3, 19-21 PLG).

Habitat: Significant at oxidation–reduction interface with the biomass, shifting from aerobic to anaerobic dominance across the front. The positioning of the plugging biomass means that nutrients tend to arrive from the reductive side and oxygen from the oxidative side. In groundwater, the plugging biomass at the interface commonly dominates bacterial growths.

Form: Biomass is composed of many bacterial communities in a generated plug. On the oxidative front, the dominant bacterial communities are commonly aerobic, often dominated by heterotrophic and iron-related bacteria (IRB). This side of the biomass may act via respiring organics (aerobically) but may also accumulate various cations such as ferric and manganic salts. Often the side of the biomass may reflect this activity when accumulated ferric ions harden to shades of brown. On the reductive side, the dominant metabolism is fermentative; the major daughter products are fatty acids and hydrogen sulfide, both of which can initiate corrosive processes.

2, 23-14 PEY (Pink Eye)

Function: Localized eye infestation, also known as contagious ophthalmia, affecting humans and livestock.

Habitat: Dominates as an eye infection; the eye and surrounding tissues take on shades of pink.

2, 16-18 PUS (Pus)

Function: Localized infestation of a tissue that exerts pressures on surrounding tissues as the infective process gains biomass. The biomass consists mainly of viscid liquids that, if not controlled, will eventually erupt through tissues along the lines of least resistance.

Habitat: Specific tissues in warm-blooded animals. The infective biomass within the tissues continues to grow. The pressures exerted on the surrounding tissues cause biomass content to evacuate as pus from the tissue.

Form: Thick, viscid, yellow or greenish liquid produced by infected tissue. Colors reflect the dominant species of the bacteria consorm. For example, dominance of green indicates that *Pseudomonas fluorescens* is likely to be one of the infective agents.

2, 06-18 ROE (Rotten Eggs)

Function: Nature prepacks avian eggs in shells to protect their contents (embryos) to the point of hatching. Eggs are rich in proteins and fats and provide the sustenance to maintain embryo growth through hatching. The fine porous nature of the shell prevents most bacteria from invading the egg. Lysozymes provide the second line of defense by destroying bacterial cells. Rotten egg consorms must be able to penetrate the shells and ward off lysozyme shields. The reward for successful penetration of an egg is a rich smorgasbord of nutrients. Relatively few bacterial consorms can effectively cause egg rot.

Habitat: Eggs are commonly incubated at blood heat to stimulate growth of the embryos. Heating also provides optimal temperatures for the activities of many mesotrophic bacteria. Infection of an egg requires penetration of the shell; penetration is more difficult if the shell remains dry and clean. Patches of organics holding water on shell exteriors and shell fractures are two portals by which bacteria can enter an egg. A series of membranes and lysozymes further retard activities. Only a few bacteria cause rotting in eggs.

Form: The three forms of rotting of eggs are : (1) dominating the egg white, causing the albumin to become very thin; (2) dominating the shell and white of the egg, generating red pigments; and (3) total invasion of the egg with structural dissolution and release of hydrogen sulfide (rotten egg odor).

2, 14-15 ROT (ROTTING)

Function: Rotting causes a living or dead organism or product of an organism to degrade structurally during infestation by microorganisms involved in bacterial consorms. It occurs as a combination of liquefaction and discoloration, often accompanied by unpleasant odors. Rotting is associated with breakdowns of cell walls and subsequent degradation of the contents.

Habitat: Usually occurs in harvested plant products under humid conditions. In animal-based food products the term "spoiled" is widely applied; the products become unusable because of more subtle color changes and odor generation from liquefaction.

Form: A loss in structural form appears in a zone that has experienced liquefaction. Carrots (*Erwinia carotovora*) are very vulnerable to rotting. Fruits and vegetables often develop similar symptoms that may be called bruising (damaged tissues leading to uncontrolled endemic enzymic activity). Physical bruising and microbiologically influenced rotting of damaged tissues are the key components to rotting events.

2, 16-21 RZZ (RHIZOSPHERE)

Function: Plant roots in soil form complex relationships with soil microorganisms via what is essentially a "barter" system. Plants must recover water and nutrients from the soil to survive and grow. Soil microorganisms benefit when plant roots bring organic (primarily photosynthesized) products along with oxygen into the soil ecosystem. A result of this complex exchange, an enhanced microbial biomass grows around and extends outward from the roots. The rhizospheres of fully grown trees may be many meters white.

Habitat: Within or near plant roots in the soil; generates a greater biomass than those generated by surrounding soils not influenced by roots. The rhizosphere is a major factor in soil fertility.

Form: Lateral growth of biomass around roots in semisaturated and highly saturated soil, particularly at the oxidative–reductive interface. Rhizospheres resemble donuts; they form rings of biomass around individual roots and around whole root systems.

2, 22-07 SHN (SHEEN)

Function: Changing reflective nature of water surfaces, generating a full spectrum (rainbow effect) of glowing colors..

Habitat: Biofilm growth at the air–water interface extending downward in the water column usually by a few millimeters. The sheen is primarily formed at the interface.

Form: Iridescent color reflecting from a water surface; originates from floating biofilms forming the interfacial biomass.

2, 23-21 SKN (Skin)

Function: Skin is commonly the outer coating of an animal. Skins are also created by biofilms growing on surfaces and interfaces. These microbial skins serve at least partially as means of accumulating chemicals such as metallic cations, carbonates, oxides, and hydroxides in a manner that gives the skin a concretious structural integrity.

Habitat: Very eutrophic environments that interface between the water and air and exhibit high salt content. The biomass may include an oxidative–reductive interface.

Form: Often a hardened surface coating that floats on or near the surface of a body of water. It may appear light grey and wrinkled. Color can range from orange to brown if significant ferric iron is present.

2, 11-16 SLT (Silt)

Function: Silt particles are smaller than sound but larger than clays. Silt is a fine sediment deposited in water. It can concrete into siltstone. Silt can form small fractions with high organic content in soils. Waters may become clouded if silt is present.

Habitat: Silt commonly contains significant organics and a bacterial consorm that accumulates fine particles within the biomass. This growth is quite slimy and may form concretious structures such as siltstones under appropriate environmental conditions.

Form: Silt is amorphogenic (no set shape); it fills voids between larger particles such as sand and co-occupies spaces within clays.

2, 21-22 SLM (Slime)

Function: Thick and slippery mud, mucous, or similar material. Slime is a biomass bounded by entrapped water and held in place by EPS. Slime provides a habitat for bacterial consorms. While laterally unstable, it provides a protective coating for incumbent microorganisms.

Habitat: Slime forms on wet surfaces commonly associated with ORP fronts. It often has direct conduits to more reductive regions in the water column bounded by fractures or porous geological structures.

Form: Slime has no shape other than that created by water flow, gravity, and positions of supporting surfaces; it is dominated by EPS-bound water. White, grey, black, orange, and brown are the most common colors, Violet slimes are rarer and reflect dominance of Janthinobacterium or Chromobacterium species under oxidative conditions or Chromatium under reductive conditions at low levels of light.

2, 24-13 TCT (Thatch)

Function: Bacterial consorms can generate thatch (a woven roof) by infesting some grasses. The bacterial biomass forms on the leaves, causing the leaves to bind together due to the hardening of EPS during growth. Thatch may become impermeable to rain by deflecting it to the side away from the thatch. Grass subject to such infestation will tend to dry out, partly because the soil under the grasses receives no local water from rainfall or irrigation due to diversion by the thatch. Such phenomena are commonly called "wind burns" but examination of the inter-woven grass reveals: (1) high levels of bacterial activity; and (2) impermeability to water. A simple test for impermeability is to pitch 10 liters of water in a bucket over a suspected site. If the water trickles over the thatch as beads and is then deflected to a healthy grass area unaffected by thatching, the infestation can be confirmed and mapped.

Habitat: Interconnected living plant material where growth in the form of bio-mass "glues" plant material (commonly turf grass) together in a thatch and prevents water penetration.

Form: Thatch infestations are more recognized by secondary impacts on plants. In turf grass, the onset of bacteriologically induced thatching leads to drying out of the grass and underpinning soil. This may be tested by coring the soil immediately after rain or irrigation. Bacteria are easily recovered from a thatch; most are aerobic slime-forming bacteria (SFB).

2, 15-13 TBD (Turbidity)

Function: Bacterial consorms growing in an aqueous environment of high clarity are likely to cause a reduction in clarity when the dispersed but growing biomass absorbs light. Turbidity may be evidenced by even cloudiness, localized cloudiness appearing as floating plates or clouds, or as buoyant particles that may maintain position or cycle up and down the water column.

Habitat: Water with nutrient content sufficient to generate a biomass that may be dispersed in the water or concentrated in EPS structures such as threads or poorly defined particles. In flowing water, turbidity is more likely generated by the shearing of biofilms attached to surfaces.

Form: Turbidity is diffused or condensed into structures that sometimes exhibit the ability to modify particle density. In hydrocarbon-rich soils, the particles are buoyed up by gas bubbles and descend after gas is exhausted from soil sampling using the SLYM BART protocol.

2, 07-20 ULC (Ulcers)

Function: Open sores on an external or internal body surface that corrupt tissues. Ulcers can also generate pus. Their corrupting influence may result directly from infestation of tissues or from an infestation associated indirectly with the ulcer and producing antagonistic effects. Generation of pus indicates the former pathway is most likely and the infestation has formed focal sites where biomass is generated.

Habitat: The gastroenteric tracts of warm-blooded animals. Stomach ulcers are associated with infective processes involving *Helicobacter pylori*. The failure to demonstrate responsibility of a single strain for other ulcerative conditions led to the classification of ulcer events as cancerous.

Form: Bacterial growths of gastric ulcers caused by *Helicobacter pylori* may be observed around the ulcerative condition. In other cases, growths may exist within ulcers. Alternatively, the growths are elsewhere and cause irritations that generate ulcers.

2, 12-07 URN (URINARY TRACT INFECTIONS)

Function: The urinary tract is considered sterile; urine is stored in a bladder until urination. The bladder and urinary tract have the potential to become infested from the kidneys, discharge ports, or the general body cavity. Urine is a rich source of nitrogenous compounds; urea dominates.

Habitat: Nutrient-rich aqueous habitat at a constant temperature but subject to the body's normal defense mechanisms that can limit primary invasion by potential pathogens.

Form: Infection begins through a combination of irritation and turbidity, particularly during discharge. Infectious biofilms are likely to form on surfaces but usually fail to cause complete plugging. Partial plugging (indicated by prolonged, slow, or painful urination) may be a common secondary feature of the infestation.

2, 12-13 WLT (WILT)

Function: Bacterial consorms (commonly in the presence of fungi) infest plant stems or leaves, blocking the phloem and xylem passageways with biomass. This causes a loss in turgor pressure inside the infested cells and leads to wilting.

Habitat: The focal site is normally the of stem just above soil level. This region is subjected to the most severe stresses created by the movement of the canopy supported by the stem. These stresses weaken the tissues, making them more likely to become infested.

Form: Infested plants lose internal pressures from fluids flowing in the phloem and xylem. This leads to wilting or drooping.

3, 10-21 BPL (BLACK PLUG LAYER)

Function: Lateral biomass may form at the oxidative–reductive interface or at the static water level in soil. When this interface or water level is relatively stable, the biomass can grow to a sufficient density laterally and plug the downward descent of water, diverting it instead to the sides away from the influence of the plug. Black plug layers are common in golf greens where conditions assure constant water levels and high nutrient loading on a suitable porous base medium (e.g., sand).

Habitat: Soils or sand greens where water and fertilizers are applied regularly to enhance plant growth. Bacteria and plant roots compete for space. On some occasions, the plant roots die back and allow the black plug layer to dominate.

Form: The layer transforms from grey or brown to intense black when a biomass grows at the oxidation–reduction interface and has an adequate nutrient base (from fertilization and destroyed plant roots) along with sulfates. The bacteria will generate hydrogen sulfide that reacts with iron to form black iron sulfides. One early symptom of the domination of black plug layer on turf grass greens is the poor condition—usually chlorosis and dieback.

3, 10-27 BBR (BLUEBERRIES)*

Function: This colorful name applies to spherical objects that litter some regions of Mars. Blueberries also occur under natural conditions on Earth, mainly in association with bioconcretious growths. Spherical balls within concretions measure a few microns in diameter. Their functions in iron-rich bioconcretions (e.g., rusticles) are unclear. The first of two classes has a waxy appearance and appears in small clusters. Iron-rich balls are dispersed throughout concretions and can cause gouging (channeling) in steel walls subjected to significant fluid flows through biofouling. This erosive channeling can cause integrity loss and leaking of pipes.

Habitat: Iron-rich bioconcretions that occur under oxidative or oxidative–reductive interface conditions in a nutrient-poor (oligotrophic) environment saturated with water.

Form: Surface scattering of blueberries on a surface indicates degradation of the bioconcretion that generated the growths. Microscopic techniques using low magnification and reflective light sources allow observation of these spheres within crystalline structures.

3, 03-16 BWR (BLACK WATER)

Function: Black water appears in reductive environments that are rich in iron and sufficient sulfur compounds and organics to generate hydrogen sulfide that reacts with the iron and generates black iron sulfide compounds. These conditions lead to a slowly growing anaerobic biomass that generates black colloidal materials in the water.

Habitat: Commonly in surface water bodies at or just above the mud line. Black water also appears in open-fractured saturated and reductive rocks.

Form: Black, odorous, viscid water sinks when it is more dense than the surrounding water.

3, 19-03 CRB (CARBONATES)

Function: Many bacterial events lead to bioaccumulations. Mountains containing dolomite may have been generated by bacterial activity! IRB BART™ analysis showed that most water samples tested exhibited white deposit in basal cones within 12 hours. The basal cones are rich in carbonates and iron-related bacterial activity may trigger the generation of these carbonates. Analysis of carbonates in groundwater samples using IRB BART led to the effective diagnosis that all biofouling involves a carbonate component best addressed by an acidic component in the treatment protocol. Carbonate (CO_3^-) is a more oxidized form than carbon dioxide (CO_2) and may therefore be generated under oxidative conditions.

Habitat: Deposits may occur in a skin on the surface of a stagnant eutrophic water body or in association with elevating carbonate deposits that generally form at or above the water line under oxidative conditions.

Form: A crystalline carbonate deposit is formed as a terminal daughter product that may or may not contain residual active viable elements of the bacterial consorm involved in the formation.

3, 05-27 CGG (CLOGGING)

Function: Obstruction of water flow in a porous or fractured medium; the clogging element contains a considerable inorganic burden. This is the reverse of plugging in which the blockage is primarily organic biomass. The biomass involved in clogging is more limited and has two major features: (1) casual entrapment and holding of sands, silts, and clays that become integrated into bioconcretions; and (2) biochemical synthesis of daughter products by the biomass. The products commonly contain ferric iron forms, carbonates, and accumulated metallic cations. Based on high inorganic content, clogging may be initially be perceived as a product of geochemical processes; this may be supported by the low levels of microbiological activity detected in drier, aged clog materials.

Habitat: Generally initiated: (1) at the ORP front where reduced chemicals move into an oxidative zone where biomass concentrates; or (2) at points of confluence, particularly in groundwater systems where the water flow transports sands, silts, and clays that become entrapped in a biomass growing within the turbulence in the groundwater. Groundwater moving toward a bore hole can accelerate from demands created by active water wells, leading to turbulence that can trigger a bacterial consorm to generate clogging types of biomass.

Form: Hardened material with low to moderate porosity. The clogging biomass has a granular structure when sand is entrapped. Color depends on materials accumulated. Browns dominate if ferric iron is present; blacks dominate if forms of manganese accumulate. In high-sulfur materials, black color may originate from iron sulfides.

3, 06-16 CRS (CORROSION)

Function: Corrosion formerly indicated the wearing away of a surface as a result of chemical action. The action caused gradual destruction of impacted surfaces such as steels and concretes. While chemical activity is certainly involved, microbiological activities also influence corrosion events. Microbacteriologically influenced corrosion (MIC) is now considered the instigator of chemical conditions produce corrosive decay. The major known routes for corrosion are (1) the generation of hydrogen sulfide from sulfates or organic sulfur containing amino acids or (2) the generation of organic or inorganic acids. Sulfate in water has long been linked to the reductive generation of hydrogen sulfide that causes electrolytic corrosion of iron alloys such as steels. The usual organic energy sources are shorter chained fatty acids. Hydrogen sulfide may also be generated reductively via degradation of sulfur amino acids. Corrosion from inorganic acids is associated with the activities of sulfur-oxidizing microorganisms that oxidize reduced forms of sulfur to a combination of sulfate and sulfuric acid. For organic acids, the short chain fatty acids (e.g., acetic and lactic) are generated in the reductive parts of the biomass and cause local pH values to drop into the mildly acidic range. For example, an ochre is a ferric-rich biomass that may be underpinned by reductive conditions that will produce modestly corrosive organic acids.

Habitat: Most events occur on the reductive side of the ORP front. Exceptions are inorganic acids generated primarily by the sulfur-oxidizing bacteria that generate very low (<3.5, highly acidic) pH conditions in an oxidative environment. Organic acid corrosion can also occur in an oxidative environment where the MIC biomass generates more reductive internal conditions, usually at the attached surfaces, leading to corrosion.

Form: Corrosion is commonly recognized by the impact of MIC and/or chemical activity on surfaces that leads to decay and /or loss of surface integrity. Two critical issues are the pitting of the surface followed by perforation when the corrosion completely penetrates the material. Pitting may also be lateral and cover a large surface area. Vertical pitting penetrates a single site as a distinct pit.

3, 18-25 CCR (CONCRETIONS)

Function: Concretion describes a hardened mass composed of several ingredients. Concrete is a synthetic concretion dominated by mixtures of gravel, sand, cement, and water. Subjecting a concretion to microbiological influences produces a natural coalescence of the hardening components over time as the material hardens to set and cure. Bioconcretions generally have complex internal crystallized structures that also create porosity in the hardened material.

Habitat: Factors influencing the formation of concretions are: (1) the need for water; and (2) location at the oxidative–reductive interface under conditions providing sufficient organics to create EPS which triggers the concreting process.

Form: The final product of a concretive process is a hardened mass of coalesced parts. The role of microbiological influences in these coalescent processes remains to be fully established. Evidence indicates that sedimentary rocks formed during periods of geological activities involved bioconcretive processes—the lateral changes seen in sedimentary rocks may relate to historical positioning of the oxidative–reductive interface during sedimentation.

3, 20-28 FRD (FERRIC-RICH DEPOSITS)

Function: Many bacterial consorms accumulate iron in oxidized forms as ferric salts. The salts build up within the biomass until the concentrations become too dense to allow growth. Iron-oxidizing bacteria appeared to play a major role in these activities. However, shifts of iron from ferrous to ferric participate in the iron shunt cycle in which oxidative conditions allow domination of the ferric forms while reductive conditions triggered by iron-reducing bacteria (IRB) give rise to ferrous forms. The IRB term covers both types of bacterial activities that may be performed in the same consorm. The generation of ferric forms of iron is exothermic and some of the energy is utilized by ribbon and sheathed iron bacteria.

Habitat: Ferric-rich deposits are formed within porous or fractured media near the static water level and where the oxidative–reductive interface forms. Conditions beneath this region are likely to be reductive while conditions above it are oxidative and only semisaturated.

Form: Deposits often resemble lateral, ferric-rich, hardened bioconcretions. If they extend along an oxidative–reductive interface, they form an iron pan (see 3, 22-18 IPN).

3, 22-18 IPN (Iron Pan)

Function: Iron pans are ferric iron-rich deposits that form as lateral layers in or just under the subsoil at a historical static water level at an oxidative–reductive interface. The pans remain as lateral ferric iron-rich strata unless eutrophic reductive conditions are created below the pan.

Habitat: In porous or fractured media associated with stable static water levels in regions underpinned by ferrous-rich groundwater.

Form: Dark brown layer in geological stratum that may show lateral unevenness reflective of historical shifting in the oxidative–reductive interface when the pan formed. A pan can cause low levels of permeability through bioconcretions, plugging the pores and fractures in the infested media. This impermeability can tighten; water will pool over the pan and form a major water source for agriculture. Deep ploughing to break up a pan can allow the perched water to pass into the groundwater, leading to arid soil.

3, 18-14 LSL (Lateral Slime Layer)

Function: Slime can form the bulk of a biomass without heavy accumulations of metallic cations under eutrophic conditions with low metal content. The slime is supported by water bound in EPS on a surface or within porous or fractured media. Most of the biomass forms across the oxidative–reductive interface and can easily dissipate or move if the front relocates.

Habitat: Lateral slime layers commonly create impermeability which reduces the transmissivity of water through the slime layer infestations.

Form: Homogeneous slimes extend laterally through the infested region, usually the differentiation point (interface) of oxidative and reductive conditions. Generally grey and translucent unless chemicals are accumulated within the slime. Slimes may be colored under oxidative conditions if the bacterial consorm includes pigmented species of bacteria or fungi.

3, 15-28 MGN (Manganese Nodules)

Function: Accumulate on the ocean floor and concentrate manganese in a hardened biomass. The growths concentrate in ocean floor regions with high content of manganese at the mud line or near vents that seep high concentrations of reductive manganous-rich waters.

Habitat: Nodules grow at or just above the oxidative–reductive interfaces on deep ocean sediments. During growth, they oxidatively accumulate mainly manganic oxides and hydroxides and become denser. This density may lead to the sinking of matured nodules into the sediment where reductive conditions may trigger the release of manganous forms that migrate upward into the oxidative environment and are again accumulated in nodules via a series of cycles.

Form: These rounded or flattened structures, when sectioned, show distinctive layering. In spherical nodules, the layers resemble annular rings that indicate routine cyclic events that control growth. Bacterial consorms can be recovered in high numbers.

3, 18-19 OCR (Ochres)

Function: These mineral incorporating mixtures of clays and ferric oxides are used to produce light yellow, brown, and red pigments. Two functions are: (1) passive accumulation of clays incorporated in the biomass; and (2) active accumulation of oxidized forms of ferric salts. An ochre appears to function actively as a complex biomass at the oxidative–reductive interface, particularly when reductive waters high in ferrous iron forms enter the growth. Ochres function as complex bacteriological consorms that accumulate ferric forms of iron as they grow.

Habitat: Interfaces where reduced forms of organics and ferrous forms of iron enter an oxidative condition. The organics are degraded and used for growth; soluble ferrous forms are oxidized to insoluble ferric forms. The biomass hardens rapidly due to crystalline ferric iron deposits.

Form: Ochre gradually hardens with growth and maturation, partly due to gradual increases in the gravimetric percentage of ferric iron (6 to 20%) with maturation to 70 to 85% when the ochre becomes incapable of functioning efficiently and becomes a spent geologic material.

3, 10-30 PGI (Pig Iron)

Function: Pig iron consists of terminal ochrous deposits in which the bacterial consorm biomass can no longer function efficiently due to accumulations of ferric iron. While pig iron is a terminal product of bacteriological activity within ochres (3, 18-19 OCR) or rusticles (3, 19-26 RST) it still contains surviving (resting) ultramicrobacteria that may become active again if environmental conditions change (e.g., dispersion of the ochre in more eutrophic conditions)

Habitat: Pig iron is found at certain points in geological time when strata at the oxidative–reductive interface were subjected to growth of ochres or rusticles. A layer of pig iron in sedimentary rock indicates a past site of an interface, presumably with oxidative water above and ferrous iron and organics flowing upward from below.

Form: Since pig iron started to collect at specific periods when environmental conditions were favorable, deposits generally tend to appear as lateral bands in sedimentary rock. These bands are usually dark brown and resemble compressed encrustations. The various forms of pig iron are categorized by composition, possibly of secondarily bioaccumulated chemicals. Silica and manganese are commonly incorporated in the iron along with organic carbon, phosphorus, and sulfur arising from bacteriological activities during formation.

3, 03-19 PFR (Perforation)

Function: Biological or chemical removal of support materials through erosive activities in a defined region that generates a hole. Bacterial consorms are involved when the primary agents for perforation are generated by corrosive daughter products from consormial activity. The product of this activity an eroded hole that passes through the material of concern (steel, aluminum, concrete). Perforation may be viewed as the end product of a process that may have involved microbiologically influenced corrosion (MIC).

Habitat: Perforation involving MIC goes through phases: (1) location of a biofilm, (2) formation of corrosive pitting, and (3) full perforation of the infested solid. After perforation, the containment material can no longer function as an effective storage barrier and stored materials will leak through the perforation.

Form: Perforation is preceded by a pitting in which the surface under attack begins to destabilize and break up. This may happen under eutrophic conditions around an oxidative–reductive interface. Daughter products that cause destabilization are the corrosive elements formed in the biomass and associated with growth. Pitting may be electrolytic or acidulolytic and may be controlled by impressing strong cathodic charges into the material. Cathodic charging reduces the potential for biofilms to attach to and grow on the protected surfaces of concern.

3, 06-24 PTG (Pitting)

Function: A biofilm attaching to and growing over a surface produces growths that cause deterioration of the surface. This deterioration involves a loss in thickness in the underpinning surface as a result of interactions of daughter products of consormial growth in the biomass. These interactions may be corrosive or electrolytic and make the surface uneven and pitted. Pitting may deepen, perforate the surface, leading to loss of integrity and leakage.

Habitat: Surface biofilms attach and cause pitting under reductive conditions generated in the biofilms.

Form: Crust-like (encrusted) surface extending downward toward a compromised surface in a series of lateral waves—sites of bioaccumulation of ferric iron, other metallic cations, and possibly calcium. Perforations may be detected by the intensity and density of the waves. Encrustation may appear concave at those sites. Pitting may also play a role in plugging early perforations (primordial holes) to prevent leakage. Multiple micro-perforations cause steel pipe to "sweat." Because of the micro scale of the perforations, they are relatively easily plugged by the biomass associated with the pitting. The net result is a slow release of water through the micro-perforations.

3, 19-21 PLG (Plugging)

Function: Plugging is the filling of a gap, cavity, or void with a biomass that acts as a wedge or stopper to prevent the passage of fluid or air. Plugging in a microbiological sense refers the generation of biomass within the gap, cavity, or void that prevents the passage of water. Biomass requires sufficient robustness and integrity to withstand the physical forces inflicted upon it by water. A younger biomass is dominated by EPS-bound water and has a gel-like consistency. As it matures, it accumulates surplus nutrients and chemicals such as metallic cations, carbonates, sands, silts, and clays. When integrated into the biomass, they improve its integrity and make it more resilient to forces inflicted by the water. Plugging therefore is slow to be recognized since the biomass repeatedly undergoes pulses during growth. The pulses involve expansion, stabilization, and secondary compression of the biomass volume. This translates into step functions that cause a loss in transmissivity followed by stable flow, then improved transmissivity as each cycle ends. This step function when related to water well production may indicate that plugging is present. Step functions can also be observed in oil and gas wells but the generation of plugging by biomass has not been validated.

Habitat: Plugging biomass usually grows at oxidative–reductive interface where the potential moves from reductive (rich in reduced nutrients) to oxidative (rich in oxygen). These fronts are commonly found in porous materials (such as sand), fractured sedimentary and igneous rocks, and upper surfaces of sediments. Plugging directly affects the transmissivity of water through pores and fractures. Plugging (commonly lateral) in sediments forms a barrier to the upward movement of gases. This may cause the gases to be perched on the reductive (lower) side of the plug. These gas voids erupt occasionally and vent the gas into surface waters.

Form: Plugs have distinct edges that appear gel-like in a young biomass and harden as the biomass matures. Older plugs may incorporate sands and clays that affect the texture. If a plug feels slippery when rubbed between two fingers, it is likely dominated by clays or EPS. If rubbing reveals granular material, the plug may contain sands and possibly ferric oxides and hydroxides if the plug accumulates iron. Water content is high in young plugs, but can fall dramatically with maturity. Treatment of an environment to remove a plugging biomass and regenerate original characteristics presents a number of challenges. Water wells are treated to regenerate productivity by removing a biomass. Many techniques such as shock chlorination produce only short-term gains (the biomass responds to the chlorine by compressing). Shock treatment reduces resistance to flow and production may return to an extent, but the reactive compression is quickly followed by a corrective expansion in the biomass after 2 to 6 weeks indicating that the effects of shock chlorination have ceased. Modern multiphase treatments involve a sequence of chemical treatments and temperature adjustments. These treatments tend to drive the biomass from surfaces in a three-phase sequence (shock, disrupt, and disperse). Many effective treatments involve the application of the three phases along with temperature adjustments. Driving the environmental temperature up at least 30°C seriously traumatizes bacteria within a biomass. The principal benefit of dropping temperatures down to freezing is the stripping of biomass forming the plugs from surfaces, but it does not kill the bacteria in the biomass (see also 2, 19-21 PLG concerning alpha two consormial plugging with high organic and bound water contents).

3, 19-26 RST (Rusticles)

Function: A rusticle is a ferric iron-rich growth formed on the insides and outsides of hulls of steel-clad sunken ships. This term was first used to describe the growths observed on the bow section of the *RMS Titanic* when the wreck site was first visited in 1985. Since then, rusticles have become common observations on steel ship wrecks in deep-ocean environments, particularly at depths exceeding 1,500 meters. These growths appear to be generated by complex bacterial consorms with highly porous biomasses structured by crystalline forms of ferric oxides and hydroxides. Their function, at least in part, is to extract the phosphorus, sulfur, and manganese from the steel as nutrients while electrolytically (or acidulolytically) extracting the iron from the ship's steel. The several forms of rusticles share a common ability to attach to steel surfaces and make use of the flow patterns of water around the wreck site. Functionality of a rusticle appears to be driven by extracting nutrients and iron from the steel in a manner that may leave elemental carbon behind—the only steel component not utilized. Rusticles cause biodeterioration and collapses of steels and other iron alloys until they are compromised, assimilated and/or dispersed.

   Habitat: Rusticle consorms are found in all submerged steel shipwrecks. Ships lying within photic zones (where light penetrates) tend to attract diverse flora and fauna in addition to the microbiota generated by microorganisms. After a rusticle biomass generates more than 20% ferric iron content, predatory fauna are repelled by the iron and the ferric-rich biomass survives. Rusticle biomass is most common in low flow waters such as those inside a hull, provided the environment remains oxidative. Rusticles are not found in reductive regions unless the oxidative–reductive interface vacillates. At the *RMS Titanic* site, a 5- to 7-year life cycle from early growth through maturation and decay was observed. Littered around the *RMS Titanic* are "spent" rusticle husks with iron content exceeding 85%; they may be classified as forms of pig iron.

   Form: Rusticles are ferric iron-rich bioconcretions that are porous internally and possess multilayered impermeable coatings that hold ducts that connect from the outside environment to a complex web of channels and interior water reservoirs. Studies indicate that the various bacterial communities are established at several focal points within rusticles; most of the volume contains little or no bacterial population activity. Shape is dictated by the environment where the rusticle grows. Exudates from the ducts contain yellow colloidal material (Fe concentration <8%) and red dust (Fe concentration 15 to 20%). This colloidal and particulate material is rich in bacterial activity and may serve as the main means of reproduction and colonization.

3, 10-10 SBL (Sand Boils)

Function: Sand boil is a civil engineering and hydrologic term that describes sand that bubbles up with water. This "boiling" is very noticeable. The sand appears unstable as the water flows upward and outward to relieve hydrostatic pressures. Sand boils are symptomatic of the failures of sand strata to relieve hydrostatic pressures in a diffuse manner. The strata under the sand are usually biofouled; the biomass occupies a significant percentage of the voids and operates as a plug, diverting the water along the lines of least resistance. If the lines extend upward through the over-bearing sand strata so that water pressures are relieved, the sand appears to boil as the water under pressure moves upward and is released. These boils can also form in association with geotextile membranes plugged with biomass. The net effect is for the membrane to swell (form a hemispheric bubble) and then find relief. The swelling or boiling has moved rocks put in place to stabilize the surface environment.

Habitat: Porous media, often with sand strata; occasionally in geotextile membranes. The primary habitats are the voids of porous media where conditions move from reductive to oxidative across the interface. Often the biomass is dominant in media saturated with water.

Form: Biomass associated with plugging voids in porous media is generally heterogeneous. The thickest growth occurs at the oxidative–reductive interface. Growing plugging biomass structures must allow the passage of water so they can extract reduced nutrients and recalcitrant materials. A biomass is not a solid harden-ing material that includes bound water; it is a very porous structure that allows the diffuse entrance and exit of water. Boils are thus products of the relief of hydrostatic pressure via the collapse of the weakest parts of a biomass that created conduits. In structures subject to biomass biofouling and plugging (e.g., levees), regenerative (rehabilitative) treatments may reduce the scale of biomass encroachment. If sand boils are common in a biofouled system, treatment must be applied to reduce or eliminate the boils, primarily by managing the generation of biomass within the sand and underpinning formations.

3, 15-17 TCL (TUBERCLES)

Function: Tubercles are surface growths that exhibit unique features and are focused sites for corrosion of steel products. A tubercle provides a relatively impermeable cap usually dominated by ferric iron within which a biomass forms, cloistered away from the outside environment. This biomass involves a consorm that triggers corro-sive processes in the underpinning steel surfaces. The processes may be electrolytic (driven by hydrogen sulfide production) or acidulolytic (driven by the fermentative production of fatty acids). In general, tubercles are usually associated with corrosion rather than plugging.

Habitat: Walls of pipes, particularly those made of iron-rich alloys. Tubercles can also be present in pipes that transmit water with high iron content. Tubercles congregate at points in pipes where dissimilar metals join, for example, welded joints, rivets, and sections made of alloys. These growths may also be more frequent along pipe areas subjected to stresses during manufacture or by natural geologically induced shearing stress.

Form: Tubercles resemble elongated nodules topped by eruptive extensions. The eruptions are coated and harden over time. On a surface, a tubercle may exhibit a slime-like biomass. When removed it, will reveal pitting at the points of contact with underlying metal surfaces.

4, 05-25 BAP (Black Asphaltene-Rich Plugs)

Function: These growths appear in dense petroleum hydrocarbons such as crude and heavy oils. These extracted hydrocarbons are highly reductive in forms with high carbon and hydrogen contents and few nutrients. Water is a relatively small but critical component of oils, usually at levels between 0.2 and 15% or more. The functionality of black asphaltene-rich plugs is based on the "mining" by bacterial consorms of water from the oil and binding it via EPS. This causes sequential deposition of asphaltenes around the biomass. The resulting water-rich, asphaltene-rich growths form on the walls of steel transmission pipes. They can decrease flow rates by plugging and also create corrosion where they are in direct contact with the steel walls of pipes.

Habitat: Transmission pipelines for crude or heavy oils, generally concentrating at anaerobically charged sites, compression and turbulence points in flowing oil pathways. The plugs incorporate aerobic bacteria into the consorm even though the conditions are perceived as very reductive. It is common to install cathodic protection on the steel walls of such pipes to prevent attachment of bacteria. These charges may be diverted and utilized for the limited bioelectrolytic conversion of some of the bound water to generate oxygen and create more oxidative conditions.

Form: Black asphaltene-rich plugs develop into black "goop" that grows gradually, causing plugging of perforations, risers in pump jacks, and deposits in oil storage tanks and transmission pipelines. They form corrosive sites at their interfaces with steel pipe walls.

4, 04-20 CDR (Carbon Dioxide Reducers)

Function: Considerable efforts are directed at the capture, storage, and sequestration of carbon dioxide in the crustal environment as a means to control its negative impacts on the surface biosphere. The radical increase in carbon-based fuels (e.g. oil and natural gas) is considered a prime factor in climate instability (global warming). Injecting carbon dioxide into groundwater-saturated geologic environments is reported to produce secondary benefits in the reduction of carbon dioxide to hydrocarbons. This led to claims of enhanced recoveries from oil and gas fields resulting from carbon dioxide injection. Essentially carbon dioxide reducers function by reducing carbon dioxide to hydrocarbons while utilizing the stripped oxygen and other nutrients and chemicals in the injection for metabolic purposes. The method generates a biomass along the injection pathway through the oxidation–reduction interface, leading down to the regenerating oil and gas reserves.

Habitat: We know little about the bacteriological reduction of carbon dioxide. In geological terms, the injected gas passes from oxidative to reductive conditions commonly with water-saturated or semi-saturated porous and fractured media.

Form: Biomass formed in response to the injection of liquids highly saturated with carbon dioxide develops in several strata reflecting a shift of ORP toward reductive, increasing salt concentrations in the deeper groundwaters, and producing an upward thermal gradient. While these environments may seem extreme for plants and animals, prokaryotic communities can adapt to these changes. Under these conditions, a number of laterally deployed biomasses participate in the reduction continuum from carbon dioxide to reduced forms as petroleum hydrocarbons and natural gases.

4, 05-17 COL (Coal)*

Function: Coal is a nutrient-rich, highly combustible, carbon-based material produced in various grades from high quality black coal down to peat (see 4, 07-19 PET). Coal seams vary in depth from very shallow to deep and appear associated with intensive prehistoric activity involving heavy depositions of organics under saturated and reductive conditions.

Habitat: Saturated reductive conditions with heavy burdens of collapsing surface biospheric organic materials. The materials degrade reductively, not to produce petroleum hydrocarbons and natural gases, but to generate elemental carbon in combustible forms as coal. Logically the generation of elemental carbon from which the hydrogen is stripped off would be considered an ultimate bioreductive event occurring deep in the geological strata. However shallow coal beds may indicate that the deposits reemerged through tectonic activity or were created near the surface. This could be the case where the organic overburden was very heavy and the biomass created by the activity generated strong internal ORP gradients. The biomass may have established extremely reductive conditions that formed the critical sites for total carbon reduction, thus producing crystalline coal.

Form: Coal bed deposits evolved from very reductive conditions created in an active biomass. Examination of coals generally shows recalcitrant nutrient loadings of phosphorus, nitrogen, and other macro- and micro-nutrients, indicating that the biomass generating the coal did not need all the nutrients. This supports the concept that coal originates, at least in part, from bacteriological activity arising from a very rich organic over-burden during the later periods of the Mesozoic era (particularly the Cretaceous). Replicating these conditions to create coal by bacteriologically influenced reductive activities would certainly prove a challenge!

4, 06-22 BNG (Biogenic Natural Gas)

Function: Methane-producing bacteria (MPB) function in very reductive saturated environments and reduce longer chained fatty acids and carbon dioxide to natural gases, mostly methane. Fatty acids arise from the fermentative breakdown of organics; carbon dioxide is a daughter product of respiration. Biogenic natural gas competes with sulfate reducing bacteria (SRB) for fatty acid substrates. The SRB dominate in less reductive environments and the biogenic natural gas wins in more reductive ORP conditions; the transitional borderline ORP value is −150 mv. MPB activity can be monitored in reductive environments in groundwaters associated with water wells. The gases usually become active in more reductive zones near bore holes where fermentative bacterial activity generates fatty acids. Methane gas released by these activities may find its way into the bore hole and into pumping well water. At greater depths, the natural gas may remain active when liquid water present and methane can be produced biogenically. In extreme environments where such activities cannot occur, active methane can be produced thermogenically.

Habitat: Biogenic natural gas consorms exist under very reductive conditions: water saturation and an over-burden of organic material degraded fermentatively to fatty acids. If the ORP value is more reductive than –150 mv, the main drivers for methane generation will be the fatty acid daughter products. However, carbon dioxide generated from respiratory, combustive, or volcanic events may also trigger reduction of carbon dioxide molecules to natural gases.

Form: Biogenic natural gas activities are likely along thermal gradients leading toward magma where the temperatures are too high for liquid phases of water to exist. Although hydrostatic pressure increases act as suppressive factors, prokaryotic bacteria continue to function. Growth would be as dispersed biomass, mostly attached to surfaces. The natural gases generated would move upward to pool in reservoirs or seep to the surface. The reservoirs may be capped by impermeable geological strata or biomass-rich plugs.

4, 03-13 BSR (Black Smokers)

Function: Black smokers are deep-ocean vents, particularly those associated ridge lines such as the Mid-Atlantic ridge. They exhaust very hot waters into the ocean at several hundreds of degrees Celsius. Such venting waters are also nutrient-rich and support fauna and microbial biomasses. Growths can be seen along the thermal gradients as the waters cool to ambient (just above freezing). Bacteriologically the biomass is complex and classified as alpha four because the flows arise from very reductive environments. Bacteriological function at this active transitional interface between very hot groundwaters and cold seawaters is complex and involves consorms from a number of alpha groupings including two, three, four, and five.

Habitat: Of all of the consorm habitats, black smokers occupy one of the most extreme: very hot water venting into a cold but geologically active deep-ocean environment. Bacterial biomass generated from discharged alpha four consorms couples with high nutrient loading. The loads consist of mixtures of natural gases, volatile petroleum hydrocarbons, oils, and potentially high concentrations of reduced chemicals such as ferrous iron. Venting in oxidative cold seawaters triggers intense aerobic activity, generating a sizable biomass along with ferric iron forms as oxides, hydroxides, sulfides (sources of black color) and carbonates. These insoluble products form growing geological structures that provide habitats for the bacterial biomass and predating fauna.

Form: The diffusive nature of the streaming hot waters of the vents indicates rapidly growing biofilms attached to the surfaces and rich biocolloidal structures floating in the rising waters. The bacterial biomass triggers intense predatory activity by the fauna (primarily shrimps and crabs) that are found in abundance around black smoker vents.

4 01-28 OIL (Oil)

Function: Oils are found in the reductive regions of the crust where overbearing organics migrate from the surface biosphere. As these organics move downward, bacteriological activities lead to the stripping of the organics carbon and hydrogen—essentially the (C_xH_y) hydrocarbons. Within the upper strata of the crust, oxygen and the major macro-nutrients (nitrogen, phosphorus, sulfur, potassium, iron, and magnesium) are removed. If the product is natural gas, biogenic natural gas consorms are involved. Oil results from the activities of oil-generating bacteria. These reductive processes generate oils containing molecules 7 to 50 carbons long, primarily as petroleum hydrocarbons in fractions of crude and heavy oils. These molecules are not easily degraded under the extremely reductive conditions present at these depths and thus become recalcitrant deposits below impermeable geological or biological structures. The deposits are usually bounded by groundwaters and recovery of some oils requires mechanical manipulation of the boundaries.

Habitat: Highly reductive environments saturated with groundwater within porous or fractured geological media. Temperatures become progressively hotter and hydrostatic pressures increase with depth. Bacteria, once adapted to these extremes, should be able to function if provided a liquid water-dominated environment. The generating biomass may be located below the accumulating oil reserves.

Form: The form and location of the oil generating biomass remains conjectural but most likely is near the movement of downward migrating organics, flow paths for groundwater, and areas where recalcitrant oil is trapped. The interfaces of groundwaters and oil may form sites for the more rapid generation of crude and heavier grad oils. Such conditions may also support the activities of biogenic methane-producing bacteria.

4, 07-19 PET (PEAT)

Function: Peat is a partially carbonized form of vegetable matter that decomposes in water under reductive conditions. It is combustible and thus a source of energy, but far less efficient than coal or oil. It is generated by an acidic fatty acid-rich product that becomes recalcitrant under the conditions in a peat bed. As long as the beds remain saturated with water below the static water level, peat does not degrade further to hydrocarbons and natural gas. Fermentative production of fatty acids is the major factor in the generation of peat. The seasonal environment (temperatures, low relative pressures, lack of water flow through the peat body) appears to favor stasis when mildly reductive conditions are generated. Peat beds possess some unusual features; the best known is the ability to preserve human bodies without excessive degradation or deformation.

Habitat: A collection of vegetative organic material that collects and fills in perched bodies of water. The environment is moderately reductive but still subjected to seasonal surface environmental effects. The water bed is dead (no obvious surface inflows or outflows).

Form: Generally peat appears to be a stable mass of partially decomposed, usually brown, vegetative material. When removed from its saturated environment and dried, peat is highly combustible but does not degrade. Peat exhibits little bacteriological activity. The pH and products generated in peat seem to inhibit growth.

5, 15-10 GHY (GAS HYDRATES)

Function: Gas hydrates (clathrates) are chemically unique in water combines with methane in a molecular ratio ($H_2O:CH_4$) of 1:8. The conditions under which gas hydrates form and hold that much methane are presently under investigation. Gas hydrates form in the deep ocean (continental shelf) and under frozen soils (tundra). In the deep ocean, the gas hydrates are found below the mud (sediment) line as frozen masses of water that hold significant methane reserves. Present estimates indicate that the methane held in gas hydrates along the continental shelves is equivalent to twice the known energy reserves of coal, oil, and gas. Two features unique to gas hydrates are: (1) frozen water (ice crystals) that remain stable up to 7°C; and (2) location in sediments, with most of the mass below the mud line. Attempts to harvest gas hydrates have encountered challenges partly because the system is fragile and destabilizes easily when attempts are made to penetrate and recover the methane.

Habitat: Gas hydrates are found under significant pressures (at 300 meters or more below ground or sea level) and appear to be stable ice-fabricated, stratified bioplatforms containing high reserves of methane. One critical aspects is the relationship of bacteriological surfactants to the growth and stability of the crystalline ice:methane complex. The bacteria may form ice as a precursor to the entrapment of methane in a complex crystalline structure. Evidence now indicates that bacteria can form ice above the normal freezing point of water.

Form: Bacteriological activity in gas hydrates is differentiable into at least three segments (capping, body, and base). Capping interfaces with the oxidative environment and likely contains alpha two bacteria that can degrade methane. The bodies of gas hydrates contain the stable elements of the bacterial biomass that maintains ice in a stable crystallized form. The bases are the bacteriological elements concerned with the reductive entrapment of generated methane within the ice body above. This definition presumes that the methane is generated within the sediment and underpinning geological strata and that the hydrate is located where methane gas moves upward through fractured and porous structures.

8 Biochemical Methods for Identification of Consorms

8.1 INTRODUCTION

Identification of bacterial consorms is a challenge for lumpers because they want to identify each bacterium within a consorm. This means that the consorm structure must be damaged to remove bacteria that are identifiable as distinct strains or communities. This atlas takes the position that the identification of a bacteriologically dominated consorm should be performed at a structural level based on its location in the environment by examining the form and location (habitat) and base the identification of the consorm primarily on those findings (see Chapter 4 with more detail in Chapter 7). This approach is general and does not attempt to identify particular strains of bacteria as important consormial components at this stage. Three steps aimed toward the identification of bacteria within a consorm allow refinement of the identification process: two biochemical methodologies outlined in this chapter and cultural methods using Bacteriological Activity Reaction Test (BART™) systems described in Chapter 9.

Biochemical methodologies are aimed at: (1) determining general activity levels of a consorm; and (2) using fatty acid methyl ester (FAME) testing of a consorm to achieve greater precision. General activity levels are determined by calculating the levels of adenosine triphosphate (ATP) in a consorm based on the direct relationship between total ATP activity in a consorm and the vitality of the biomass. Section 8.2 addresses methodologies to determine ATP. Extraction and enrichment procedures for FAME analysis are addressed in Section 8.3.

8.2 DETERMINATION OF CONSORMIAL ACTIVITY BY ATP ANALYSIS

ATP is the principal molecule that stores high energy within phosphate bonds. ATP is universal in all living cells and performs that function. When cells are metabolically more active then the ATP concentration tends to rise. Concentration of ATP is measured in picograms per gram. Dormant cells have virtually no ATP while active cells generate concentrations of ATP in relationship to their activity level. Thus testing for bacteriological activity in environmental samples can commonly be achieved in a semiquantitative manner.

Methodologies for the detection of ATP focus on the ability of the luciferase enzyme to break down ATP quickly, generating light via the breakdown of high-energy phosphate bonds. The more light generated, the more phosphate bonds broken down. The initial sources for luciferase enzyme were fireflies (*Photinus pyralis*) or bioluminescent bacteria (*Photobacterium*). Because ATP testing can be conducted very quickly, it serves as a "gold standard" for the detection of biological activity.

Bioluminescence as a microbiological activity detection tool was proposed in 1968 for use in water and was commonly used in foods by 1970. Since then ATP techniques have replaced the traditional spreadplate techniques that focus on bacteriological activity levels in the sample of interest. Luciferase testing for bacterial activity has undergone significant precision improvements to make it fully quantitative. The assay measures the amount of ATP in a sample as relative light units (RLU), also called relative luciferase units) that directly relate to the viable active bacterial population. ATP has now also been adapted to detect the presence or absence of bacteria on surfaces and detect the number of active cells. This approach does not necessarily differentiate the source of the ATP activity as prokaryotic (bacterial) or eukaryotic (higher organism) cells and lacks precision in evaluating specific groups of microorganisms. It can be used to confirm activity within a sample of a consorm semiquantitatively. The RLU measures the amount of light emitted during interactions of luciferase and ATP in the presence of oxygen. This is summarized in the equation below:

$$Luciferine + ATP + O_2 \xrightarrow[Mg^+]{Luciferase} Luciferine\ (oxidized) + AMP + CO_2 + Light$$

ATP detection can be accomplished using the second-generation Quench Gone™ (QGA) test kit available from LuminUltra Technologies (Fredericton, New Brunswick, Canada, www.luminultra.com). The method uses a luminometer to measure light produced. The Luminase (luciferase) solution should be refrigerator between uses. UltraLute is used for dilution of samples. Ultralyse 7 is needed for each test.

Calibration of the Luminase before starting a test set is important because luciferase weakens over time. To calibrate, 2 drops (100 µl) of Luminase are added to 2 drops (100 µl) of UltraCheck 1 in a small (12 × 55 mm) assay tube, mixed gently, and immediately inserted into the powered-up luminometer and the "enter" button is pressed. After 10 seconds, the screen displays a calibration value (RLU_{CL}). If the value is below 5,000 then the Luminase is spent and a new calibration with a fresh bottle is required. If the number exceeds 5,000, it should be recorded for use in calculating the ATP.

For liquid samples (for suspensions of solid samples, 5 or 10% dilution is recommended) the technique is:

(1) For each sample, add 1 ml Ultralyse 7 into a 17 × 100 mm extraction tube (Tube A), 9 ml UltraLute into a 17 × 100 mm dilution tube (Tube B), and 2 drops (100 µl) Luminase into a 12 × 55 mm assay tube (Tube C).
(2) Thoroughly mix the sample, then add 1 ml to tube A with the Ultralyse. Cap the tube and mix thoroughly.
(3) Allow sample to sit at least 5 minutes to allow solids to settle.

(4) Carefully remove 1 ml of supernatant and dispense into Tube B. Extra caution is required at this point to ensure that any sediment at the bottom of the tube not be disturbed because it will produce an anomalous reading. Cap the dilution tube and mix thoroughly.

(5) Transfer 100 µl of diluted sample to Tube C and immediately place assay tube into the luminometer. Press "enter" and record the RLU_I value displayed after 10 seconds

(5) Convert RLU_I to total ATP_I in picograms per milliliter (pg/ml) with the formula:

$$Total\ ATP\ (pg/ml) = (RLU_I/RLU_{CI}) \times 20{,}000$$

When calculating total ATP as pg/ml (liquid sample) or pg/g (solid sample), total ATP must be corrected for any dilution factors used in preparation of the sample. After correction, the normal range of ATP found in bacteriologically active samples is from a low of 250 to a high of 250,000 pg/ml or pg/g:

Total ATP Value (pg)	Activity Level
<200	Virtually inactive
<2,000	Relatively inactive
5,000 to 2,000	Active
>20,000	Very active

In the event of a "virtually none" or "relatively inactive" result, enhancement of potential metabolic activity through enrichment may be required. The enrichment technique is designed to determine whether the potential ATP activity is greater if the consormial sample is stimulated under agitated and static states in sequence. This technique is known as rapid agitation, static incubation (RASI), and described below.

8.3 RASI PROTOCOL FOR DETERMINING POTENTIAL ATP ACTIVITY

RASI is a two-stage procedure designed to utilize a specific laboratory formatted BART system to trigger a cultural environment that stimulates and enriches bacteria growth within a consormial sample. Five of the six alpha groups of bacteria five have defined BART types for the RASI procedure. These are listed and the recommended tester follows on each of the next lines:

Consorm	Tester
Alpha one	SLYM BART (S)
Alpha two	SLYM BART (S)
Alpha three	IRB BART (I)
Alpha four	SRB BART (R)
Alpha five	SLYM BART (S)

In the RASI protocol, the letter in parentheses represents the type of BART. For example, A RASI protocol starting with R would indicate an attempt to trigger the cultural activities of alpha four consorms with SRB BART. See Chapter 9 for more details about BART.

RASI has two distinct phases: (1) rotatory agitation; and (2) static incubation. In the first, the BART is clipped to a rotatory shaker. When activated with a changed sample (incorporating 15 ml liquid), each rotation causes the BART ball to move from one end of the tester to the other and back again. This causes: (1) saturation of the liquid sample with headspace oxygen, breaking up any suspendable particles to expose dormant bacterial cells; and (2) dispersal of the selective cultural nutrient medium into the liquid phase from the basal cone of the tester where it originally was placed as a dried sterile crystalline deposit. Rapid agitation is routinely performed at room temperature ($22 \pm 2°C$) for 4 hours. The agitation equipment recommended is a Labquake™ 3,625,485 shaker (Thermo Scientific, Dubuque, Iowa) operating at a speed of 9 rpm.

Static incubation varies in duration and the temperature applied, It is designed to allow the bacteria positively impacted by the rapid agitation to grow under the variety of environmental conditions created by the vertical BART. Critical to static incubation is the respective top-to-base generation of an oxidation–reduction gradient. Bacteria triggered by reductive environments are more likely to grow just above the base; aerobic bacteria are likely to be active at the interface and above.

To determine whether ATP activity was generated during RASI, the general technique at the end of the static incubation period is to use vigorous wrist action to shake the tester for 1 minute. This mixes all the bacterial growths into homogeneous suspensions. A midpoint sample should be taken within 10 seconds of the end of shaking. The potential total ATP (ATP_P) may then be measured via the standard ATP procedure.

Using RASI to measure ATP_P also makes it possible to determine whether a consormial sample had significant latent metabolic activity that was triggered. To determine the latent potential for bacterial activity in a consormial sample two total ATP determinations are required: (1) initial ATP taken from the sample without enrichment (ATP_I); and (2) ATP potential measured after completion of RASI (ATP_P). Latent ATP (ATP_L) is calculated as a percentage increase or decrease from ATP_I. An increase in total ATP indicates that RASI enriched and stimulated the bacterial consorm. A decrease means that RASI had a suppressive effect on the incumbent bacteria in the consorm. Calculation of ATP_L as a positive or negative (stimulation or suppression) percentile shift may be performed using the following equation:

$$ATP_L = (((ATP_P/ATP_I) - 1) \times 100)$$

Latent ATP_L is shown as a percentage of the initial ATP_I in which a negative percentile value would indicate that RASI enrichment failed to stimulate the activities of the bacteria in the consorm. A positive percentile value means that RASI stimulated the bacteria to produce more ATP. Essentially a negative ATP_L would be significant in a range from -30 to -100% and would mean that RASI failed to enrich the bacterial activities under those particular conditions. This does not mean that a bacterial consorm could not be enhanced with RASI; it means that the conditions for stimulating bacterial activity were incorrect. Conditions that may vary during the rapid agitation and static incubation phases of RASI are times and incubation temperatures. Further, the selection of the BART may be at fault. Selection of time lengths and incubation temperatures beyond the standard RASI are discussed in Section 8.3.

ATP_L values between −30 and +30% may be considered to be null effects: the RASI enrichment process failed to stimulate any sections of the bacterial consorm, indicating that RASI enrichment failed to impact the levels of ATP activity. If ATP_L exceeds +30%, it may be considered that RASI did stimulate the global level of ATP activity in the sampled consorm. The level of ATP_L stimulation could be determined semi-quantitatively using these guidelines:

ATP_L Value	RASI Stimulation Level
+30 to +100%	Moderate
+101 to +250%	Good
+251 to +500%	Very good
>+500%	Excellent

For the further determination of the bacteriological composition of a sampled consorm using the RASI-MIDI system (Section 8.3.2 below) and/or BART (Chapter 9), an ATP_L at least +100% should be achieved to ensure significant activity in the RASI-enriched process.

8.3.1 DIFFERENTIATION OF RASI PROCEDURES

RASI methodologies involve a number of parts summarized in the RASI code. As an example, in code WXYYZZ, W is the code for the BART employed in the RASI (S, I, or R) and defines the alpha group to be investigated (S = alpha one, two, or five; I = alpha three; R = alpha four). X defines the duration in hours (1 to 9) of the rapid agitation phase of RASI; 4 hours is the accepted standard. YY defines the employed temperature (0 to 99 Celsius) of the second phase (static incubation) of RASI and ZZ indicates hours of static incubation. The standard RASI protocol for the RASI-MIDI identification of bacterial consorms using the fatty acid methyl ester (FAME) chromatographic technique is S43020. The RASI S43020 protocol involves SLYM BART charged with a total of 15 ml of sample, agitation for 4 hours at 9 rpm, followed by static incubation at 30°C for 20 hours. S43020 translates to SLYM BART agitated for 4 hours followed by static incubation at 30°C for 20 hours.

A number of modifications of standard RASI may be made based on characteristics of the alpha group examined. Secondary considerations relate to the natural temperature at which the consorm functions. The following list of modified RASI protocols reflects these factors:

Cold-loving (psychrotrophic) alphas one, two, and five	S41272
Mesotrophic alphas one, two, and five	S43020
Warm-blooded alpha two parasites	S43720

Thermotrophic alphas two and five	S45540
Cold-loving (psychrotrophic) alpha three	I61296
Mesotrophic alpha three	I63072
Thermotrophic alpha three	I45560
Cold-loving (psychrotrophic) alpha four	R41296
Mesotrophic alpha four	R43072
Thermotrophic alpha four	R45596

These recommended RASI protocols reflect the determination of latent ATP (ATP_L) in a bacterial consorm. Potentially active consorms should generate positive percentile values exceeding +30% if the RASI protocol triggered such activity. It should be noted that RASI S43020 is the standard protocol adopted for microbiological identification schemes using FAME. This is discussed in detail below.

8.3.2 RASI-MIDI Protocols for Identifying Bacterial Consorms

MIDI (Newark, Delaware) developed a biochemical technique for identifying microbial isolates through the analysis of variations in the fatty acid compositions of their cells. Refined techniques have been developed for pure cultures of bacteria by analyzing specific "fingerprints" (chromatographs) to measure the FAME in each isolate. The use of five to twenty carbon chain lengths produces unique, replicable, and identifiable profiles as signatures of a particular isolate. FAME has become a common identification tool, particularly for health-related bacterial pathogens. MIDI has developed libraries to identify bacterial strains cultured under very specific conditions.

Identification of the bacterial consorms is performed using MIDI's Sherlock™ Version 6.0 microbial identification system (MIS) described as MIS whole cell fatty acid analysis by gas chromatography Sherlock uses Instant FAME sample preparation and Agilent GC 6850 series gas chromatographs to yield precise and reproducible FAME profiles.

Recovery of bacterial cellular growth from a tester subjected to the RASI protocol utilizes a standard Millipore filtration apparatus with a sterile 0.45 µm pore size membrane. First, moisten the membrane filter by adding 10 ml sterile water and apply vacuum of 20 psi to allow water to pass through the membrane. Loosen the tester cap but do not remove it completely. Carefully tilt the BART to pour the entire content (15 ml) subjected to full RASI into the filter membrane using the cap to prevent the BART ball from touching the filter membrane. If the RASI cultured BART develops a dense and gelatinous basal mass, gently shake the tester to homogenize the incubated content and filter only the first two thirds of the volume (approximately 10 ml). The surface of the membrane filter contains filtered cellular material from the RASI cultured sample.

Final sample preparation for standard rapid MIDI bacteriological identification requires recovery of the filtered cultured material from the membrane filter. Use a sterile plastic disposable 1 µl loop to scrape the surface of the membrane to recover entrapped cellular material (biomass from the cultured consorm). The general recommendation is to make three slow strokes with the loop flat down across the biomass-coated surface of the membrane followed by three more strokes at right angles to the first strokes. These sequenced strokes can be repeated until a clear collection of cellular materials from the entrapped biomass rises from the center of the loop. Two to five sets of strokes across a membrane are normally be necessary to collect required materials and the loop should contain 2.5 to 3 mg of cellular material on the loop—enough for MIDI testing.

To conduct MIDI testing through generating a FAME chromatograph, place the loop tip loaded with cultured cellular material into the bottom of a standard 1 ml glass vial (recommended by MIDI). To maximize the deposition of gathered cellular material into the vial, flick the loop inside the vial to force the transfer of sampled material into it. Do this until no visible cellular material remains in the loop. The next step is standard Instant FAME preparation of the cultured sample material (MIDI Version 3.3). Extraction of FAME involves three reagents used sequentially: (1) application of a mixture of methanol with potassium hydroxide; (2) solvent extraction with hexane; and (3) treatment with dilute hydrochloric acid containing a red indicator dye.

This detection method automatically quantifies the fatty acids extracted from the RASI S43020 cultured samples using the Sherlock software then used for the interpretation of the fatty acid peak profiles generated. Sherlock Version 6.0 includes the interpretation of the data as two-dimensional plots or histograms, or accurate root neighbor joining techniques without interpretation. Interpretation is normally provided by standard libraries generated via Sherlock library generation software (Version 6.0). These libraries are based on specific recommended cultural practices utilizing standard selective agar culture media. FAME data generated by the RASI S43020 technique must be applied to a unique library generated for relatable samples using the software tools included in the library generation system.

The method described here uses unique libraries that are inherently different from standard cultural practices in two important aspects: (1) the method uses broth-based techniques incorporating SLYM BART; and (2) RASI S43020 stimulates selected bacterial consorms on the premise that their fatty acid configurations will remain consistent even if the consorms contain many active individual strains.

The conventional comparison of unknown cultures with a standard library identification system commonly uses probability percentages to express the closeness of a match of unknown and established library data. The probability comparison of the Sherlock system employs a similarity index (SI). The SI number indicates how closely the FAME of the unknown compares with the mean fatty acid compositions of recognized members of the selected library. Samples with SI values of at least 0.500 are considered good matches provided the next closest identification is separated by at least 0.100 SI units. SI is generated on the assumption that microorganism strains all exhibit normal Gaussian distributions (bell-shaped curves).

RASI-enriched bacterial populations are recognized by the traits indicated by the FAME percentages generated. SI calculates similarity as the closeness of fit between the generated fatty acid percentages and those stored in the library for comparative purposes. An SI match of 1.000 indicates total similarity; 0.700 shows a three standard deviation window from the mean within which the match would fit. In practice, using the MIDI Sherlock system with the RASI S43020 enrichment method generates a close match between the unknown and the selected library (SI >0.850).

One challenge in developing a consormial identification system is that the techniques are relatively new and continue to undergo validation. The RASI-MIDI protocols and experimental libraries are still under construction. However, MIDI makes it possible to develop unique libraries using the Sherlock software and achieve rapid, more detailed analysis of bacterial consorms.

9 Identifying Bacterial Consorms Using BART

9.1 INTRODUCTION

Identification of bacterial consorms requires three levels of investigation. The first is determining significant biological activities. This addressed biochemically via exploratory determination of ATP activity (Chapter 8). The second is qualitatively examination of the bacterial consorm. The use of the RASI S43020-MIDI protocol designed for mesotrophic aerobic heterotrophic bacteria was discussed in Chapter 8. This chapter defines the third and final phase: use of the Biological Activity Reaction Test (BART™) cultural technique to achieve a quantitative and more qualitative understanding of active communities within a sampled consorm.

Unfortunately, bacterial consorms in nature cannot be simply defined qualitatively via biochemically based chromatographic techniques. These methods are precise and reliable, but they do not address the compositions of bacteria of concern within a consorm. This chapter simple differential cultural BART techniques to determine the compositions of bacterial consorms based on major bacterial communities that may be active in the consorm.

Developing any diagnostic management strategy requires that you "know your enemy" but you must also "know who your friends are." The vast bacteriological biosphere is within us, beneath us in groundwaters, surrounding us in the oceans, and above us in the clouds. We must get to know our friends because they are frequently at war with the bacterial enemies we must understand and manage.

BART can determine both the "good guys" and the "bad guys." Only when these relationships are recognized can a management strategy proceed. Humans are preoccupied with the control of disease and the negative effects of bacteria have on our engineered comfortable lives. Diseases are usually considered the products of a single microorganism (pathogen) that penetrates the body and causes significant malfunctioning. Clearly the pathogen is the bad guy and the bacteria that normally function within the body and environment are the good guys. We have created engineered systems to enhance our routines, comforts, and convenience. Some of the enemies are bacteria that throw wrenches into engineered systems; they are consorms, not single strains of microorganism.

Failures are created by many factors. A robust, resourceful consormial biomass may cause plugging; generate electrolytic effects causing corrosion; and produce unpleasant daughter products such as odors, colors, slimes and even carcinogens. These biofouling effects in engineered systems may be controlled to an extent by rehabilitative and preventative maintenance practices by treating the effects as if they are diseases. Clinical concepts dictate that diseases are produced by single strains of microorganisms and we apply the same concepts to "diseases" of engineered systems. This means an immediate hunt for the guilty microbial strain considered the enemy (pathogen) of the engineered system. The traditional rule for treating biofouling in an engineered system is blame a single species of bacteria (the pathogenic enemy). In reality, the blame may be placed more correctly on the activities of a bacterial consorm!

Historically this entrenched mindset (a single strain of bacteria is the enemy causing disease in humans and malfunction in engineered systems) leads us to diminish the roles of normal microorganisms in the protection of a body or system from acute catastrophic or chronic failures. We continue to see a problem as the product of a single strain of microorganism and the solution is to kill all of the microorganisms (both friends and enemies) at the site of impact. This mindset has limited our ability to determine the causes of failures in both living and nonliving systems. In medicine, the reliance on agar substrates to selectively culture pathogenic bacteria means that pathogens that do not grow on agar media are considered insignificant; the patient's disease may be considered a cancer simply because no bacterial agent was identified.

The effectiveness of BART in a multiplicity of environments allows the selective culture of bacteria at various points of testing. Adenosine triphosphate (ATP) can detect generalized metabolic activities. RASI-MIDI technologies allow chromatographic FAME resolution of any bacterial consorms present. When applied to sampled tissues that may contain consorms, it may be possible with these techniques to detect another stratum of (unculturable) bacterial pathogens. These enemies would have the common characteristics of inability to grow on agar and the ability to generate similar symptoms when introduced into susceptible host tissues using Koch's postulates. The view of engineered systems must change from the premise that enemy is a single strain of microorganism to the concept that a designed system will foul and/or fail because of: (1) lack of recognition of the associated biomass activity; and (2) insufficient design input to minimalize the potential for failures.

Diagnostic microbiologists must accept the idea that diseases and system failures may not always be attributable to single strains of microorganism and that bacterial consorms also cause these problems. BART is the third stage in the diagnostic process that focuses on the activities (determined by time lapse) and the types of bacteria that are active in a sample (determined by reaction patterns). BART is a simple cultural tool useful in the field, laboratory, or office.

9.2 DEVELOPMENT OF BART TO DETERMINE BACTERIAL ACTIVITY

Bart developed frustration: no simple bacterial detection tests could be employed in the field on water samples; samples had to be brought to a laboratory for testing. BART was conceived in 1986 and patent protected and trademarked. One major difference between BART and other microbiological tests, in addition to its simplicity, is that it handles a greater range of environments than other cultural detection systems. One environment is an oxidation–reduction gradient arising from the activities of microbes in a sample. The reductive side of this front lies under the oxidative side and the front moves as the oxygen in the sample is consumed. The bottom section of the BART apparatus contains selective chemicals that diffuse upward as they dissolve. This creates a diffusion gradient that forms too slowly to traumatize or harm the bacteria and encourages them to become active.

Another important feature is the BART ball (floating intercedent device) that floats on a prepared liquid sample. Its position restricts the rate of oxygen penetration downward from the headspace into the culturing fluids of the sample by creating a restricting throat-to-oxygen movement. This causes an oxidative–reductive interface to form and remain stable. It slowly rises upward as the oxygen is consumed by the bacteria at rates exceeding the diffusion rate around the BART ball. In concept, visible changes of BART determine activities (growth and production of bubbles) and reactions (color changes). BART does not demonstrate total cell population; it determines activity levels of bacteria that respond to the specific BART type.

Estimating the selected bacterial activity level within a consorm includes a very important principle: to determine only active cells (functioning in the charged tester) and not dormant (sleeping) cells. Simply put, active cells in a bacterial consorm are probably causing the problem! Activity and reactions are measured by the time lapse (TL) between setup and the first visible reaction or activity that is defined in the protocol for the specific tester. TL can be measured in seconds, hours, and days. Result times may be shorter than 3,000 seconds (HAB BART in primary influent wastewater) or as long as 15 days (SRB BART in a well with a deeply set, relatively inactive anaerobic SRB community). TL defines activity or aggressivity levels of a bacterial community generally as high, moderate (medium), low, and background. Few natural water samples are truly sterile (zero activity; no reaction or activity observed). Most samples contain at least small numbers of cells that may not be very active.

Another major factor affecting the detection of consorms in a sample is the temperature at which BART is incubated. Bacterial consorms generally function in narrow temperature ranges and bacterial communities within the consorm must also function in that range. In attempting to identify consormial activities, incubation temperatures must be kept close to the original environmental conditions in which the consorm was active. Convenience dictates that room temperature is appropriate (we all like 21 to 25°C; why shouldn't bacteria?). Many bacteria grow optimally at 27 to 30°C. For this reason, 28 ± 1°C is often selected for BART when precision and speed are important considerations. Standard conversion tables (www.dbi.ca) convert TL (depending on incubation temperature) to activity or aggressivity level and predicted population.

It is possible to predict the active bacterial population operating in a sample on the assumption that only active cells are included in the detection process (and dormant cells excluded because they remain passive during the test). Populations are expressed as predicted active cells per milliliter (pac/ml) of the bacterial community active in the BART when the sample under examination was incubated.

Two components are used in BART. The tester is essential: the sample test is performed in the tester. Precision is handled by the Virtual Bart Reader 72 (VBR72) system digitally and routinely records the sequences in which reactions and activities occur. This allows an operator to automatically determine TL. The two components are described below.

The tester uses 15 ml of sample to provide an accuracy limit at 67 cells per liter where a wholly liquid sample is used. For a semisolid or solid sample, the dilution commonly ranges from 10% (1.5 g) to 1% (0.15 g) and reduces sensitivity proportionately. The tester consists of a polystyrene tube, a loosely screwed polyurethane cap, and a dried cultural medium specific for the bacterial group monitored.

VBR72 utilizes digital video recording (regular room lighting conditions) and a USB (2.0)-compatible Cannon camera coupled to software that: (1) allows a computer to receive, save, interpret, and display activities in any tester in one of 72 rack positions; (2) records images in a manner that allows an operator through software determine TLs from the observed activities and reactions within the observed tester; and (3) converts TLs of all reactions observed to population data and archives the data. The standard VBR72 monitors up to 72 testers in customized racks at the same time.

9.3   BART SETUP

Testers come in field and laboratory versions. The field tester includes a protective outer vial that prevents odors from escaping; it also samples. The vial collects a sample, then protects the inner tester and operator from odors produced by bacterial activities. The laboratory version of BART has no outer vial and is more economical. The laboratory system generates the same reactions and activities as the field tester, but occupies less space and costs less.

BART setup involves a number of steps explained in the protocols. Essentially, the first step is adding 15 ml (final volume after dilution if required) of a sample to the inner test vial. This triggers the patented BART ball to float upward to restrict oxygen entry from the headspace into the sample. The test begins as soon as the sample is added. Do not shake or disturb the vial to ensure that an oxidation–reduction gradient will form within the culturing fluids. These fluids consist of the sample, dilution water, chemicals diffusing up from the basal selective chemical medium, and oxygen diffusing downward around the BART ball. It is important to keep the tester in a location where it will not be disturbed and out of direct sunlight. Room temperature (22 ± 2°C), 28 ± 1°C, 37 ± 1°C, or 12 ± 2°C ranges are used, depending on the temperature of the sample when taken. Regular room lighting does not greatly influence the rates of the activities and reactions that may occur in a tester. At regular intervals, depending on the incubation temperature, the tester should be visually examined to determine whether any reaction or activity has occurred is no VBR72 system is employed. Recommended examination frequencies are: room temperature, 28 ± 1°C, daily; 37 ± 1°C, twice daily; 12 ± 2°C, daily. TL represents time between the start of the text and the first observation of the activity or reaction; it is expressed in seconds, hours, or days.

BART can function at many levels in the field and in the laboratory (along with the sophisticated VBR72 system with customized software). Each tester has unique features, described below. Combination systems (tester and VBR72 with appropriate software) offer a different approach to monitoring microbial events for diagnostic purposes and are more comprehensive, economical, faster, and more convenient than traditional cultural methods or the new genomic and biochemical tests. In summary, BART allows precise determination of activities based on time lapse and fast recognition of appropriate bacteriologically influenced reactions.

Historically, BART principles can be traced back to the 1880s when Winogradski discovered a unique event: different microorganisms grew at different lateral positions within a water column inside a glass measuring cylinder placed in a north-facing window. This concept is still a standard of bacteriology training. BART builds on Winogradski's findings by determining that microbial growth in part resulted from the development of opposing selective nutrient gradients that rise, descend, and then stabilize within an oxidation–reduction interface. Different communities of microbes in samples located at different sites to become active and grow within culturing fluids in the testers. BART testers are used mostly in the water, chemical, oil, and gas industries. The main differences between BART and classical bacteriological techniques are:

1. BART testers encourage all bacteria active within a sample to attempt to grow in one or more selective environments. Classical bacteriological techniques differentiate at the species or taxon level and exclude many organisms in a sample because they unculturable via traditional approaches.
2. BART examines the activity levels of cells while classical techniques (based on agar media) automatically become highly selective, usually detecting only a limited number of species, not entire ranges of active microbial communities.
3. BART systems include the VBR72 that generates an interpretable data stream that can be stored in a computer for confirmatory and archiving purposes. The VBR72 can record the activities and reactions of a variety of BART testers at the same time. Data is stored in read-only files that may be viewed for analysis but not altered.

Traditional bacteriological thinking focuses on species and strain levels with little attention to the often complex community interactions of consormial activities. BART is based on reactions of bacterial communities rather than on single species. Little is known of the roles (protective or infective) of these bacterial communities and consorms in warm-blooded animals (including humans) because of the traditional focus at species level. This selectivity resulted in the failure to determine the powerful roles of bacterial communities in the control of and recovery from significant infections in both human beings and water wells. We can also argue that microbial communities play a major role in the transmission of virus particles between infection sites in different hosts. This area of bacteriology has not been explored and BART provides a simple method to detect these communities and their potentially important roles in health, disease, fertility, and survival.

The tester generally takes less than 3 minutes of technologist bench time to set up; more time is required to check for reactions and activities. These checks become automatic and archivable using the VBR72 system. Data may be stored for future reference as required. Timed sampling with BART testing can determine more precisely the potential challenges created by bacterial communities and/or consorms. BART offers a different approach to monitoring microbial events for diagnostic purposes, is far more comprehensive than traditional methods and more economical, faster and convenient than the recent biochemical and genomic methods.

Sections 9.4 through 9.11 detail the various tester types. They are listed by the color of the cap employed, the bacterial consorms detected, and the name of the specific tester.

9.4 RED CAP: IRON-RELATED BACTERIA (IRB BART)

IRB BART testers are more complex than most because of the numbers of activities and reactions that can occur. The iron-related bacteria (IRB) accumulate iron within colloidal or colonial growths that significantly exceeds their immediate metabolic requirements for iron. Commonly these accumulations are in crystallized forms of ferric iron as oxides, hydroxides, and sometimes carbonates. Growth generally takes on the rust color of the ferric accumulates. Hence their name reflects their ability to accumulate iron and qualitatively describes the various forms of bacteria that can accumulate ferric iron and also dissolve iron under more reductive conditions. By definition, iron-related bacteria incorporate:

> All bacteria that are able to accumulate iron in any form within the environmental matrix where they actively function.

This brief definition encompasses a number of major bacterial groups that are known but may not be associated within a common consormial grouping. These groups are described separately and their role within IRB group defined primarily in relation to the oxidation–reduction potential (ORP) of the environment within which they are active. The groups are generally active in specific parts of the ORP gradient and can attach to surfaces via the generation of biofilms, encrustations, nodules, and other forms of growth. The IRB groups are:

Iron oxidizing or ferrous-oxidizing bacteria (FOB) accumulate ferric forms of oxides and hydroxides within a growing biomass. This accumulation continues until the iron content reaches 40 to 95% of dried weight. Accumulation usually occurs at the interface of oxidative and reductive conditions, usually with ORP values between +10 and –50 mv. These bacteria may participate in the biological formation of iron ochre.

Iron-reducing or ferric-reducing bacteria (FRB) under various reductive (anaerobic or oxygen-free) conditions reduce ferric forms of iron to more soluble ferrous forms that move via diffusive and biocolloidal processes from the environment where they were created. FRB are usually found in reductive conditions where ORP ranges from 0 to –150mv.

Sheathed iron bacteria (SIB) live at least part of their activity cycle within a slime-like sheath onto which ferric iron may or may not be accumulated. These sheaths usually are formed in the more oxidative upper part of the biomass and frequently extend into the water flow. SIB move into sheaths for protection and growth and out of the sheaths for colonization in accordance with life cycle stage. ORP values generally range from +5 to +150mv for the sheathed bacteria to become significant parts of the biomass.

Ribbon iron bacteria (RIB) can extrude a single slime-like ribbon out of a cell in a specific direction from a single point. This ribbon is rich in ferric iron. The energy release from the oxidation of ferrous to ferric iron may meet a significant part of the total energy requirement. Generally these bacteria (dominated by genus Gallionella) grow on the surface of a growing biomass and the ribbons extend outward into the free water flow. The ribbons may follow Archimedean screw principles to move nutrients from the water to cells. When mature, the ribbons break off and move into the water flow where they are easy to observe microscopically. The presence of ribbons in a water sample does not mean that cells of Gallionella are present; it simply means that Gallionella growth occurred upstream and shed the ribbons. Gallionella grow under oxidative conditions (>+50 mv) over very sharp redox fronts in a biomass that make ferrous iron available to cells. This ferrous iron may originate in water flowing over the biomass in ferric form, possibly taken up and reduced to the ferrous form or it may have been ferrous iron that diffused from the deeper reductive parts of the biomass.

9.4.1 Qualitative Interpretation of IRB Reaction Patterns

IRB BART reaction patterns follow a sequence that indicates many characteristics of the IRB within a sample. The two major pathways are oxidative (CL) and reductive (FO). A sequence of reactions observable in IRB BART occur in various combinations. Reactions depend mainly on which IRB microbes are active within the sample tested. The "microbe" term is used here because fungi and possibly protozoa and algae may also influence reaction sequences.

The four chronological sequences in which the reactions can occur are divided into groups. A common first reaction is the development of a white base (WB, first sequence) forming as a white deposit of carbonates in the base of the tester. Usually this occurs within 12 hours of the test start (sequence one) and has traditionally not been recognized as a reaction in the interpretation of IRB activities in a sample but it is important in determining the treatment of biofouling in wells, cooling towers, and water treatment systems. WB formation indicates that carbonates are accumulating and any rehabilitation should take into account the need to disrupt carbonate-rich encrustations.

The first major diagnostic event (second sequence; see Table 9.1), usually within 1 to 5 days, is the generation of a cloudy (CL) or foam (FO) reaction. CL occurs more commonly when a sample comes from an oxidative region or at the oxidation–reduction interface between oxidative and reductive conditions. FO and CL reactions can occur together, i.e., microbes in a sample are mixtures of cells capable of acting both oxidatively (through respiration) and reductively (through fermentation). If the FO precedes the CL, the sample is more dominated by fermentors; respirers dominate if the CL precedes the FO. Sequence 3 reactions occur in an order reflecting bacterial dominance. Sequence 4 is a terminal reaction that occurs only when significant enteric and pseudomonad bacteria are present.

The location of any specific reaction type in IRB BART yields information about the active bacterial community in the sample. It also determine the ORP state where the sample was taken. Reactions can be interpreted using BART SOFT; they are summarized below:

CL	Heterotrophic bacteria present
FO	Anaerobic bacterial activity detected

TABLE 9.1
IRB BART Reaction Pattern Sequences

Sequence	Reaction Pattern	Comments
1	White base (WB)	Occurs more commonly in slightly alkaline samples
2	Clouding (CL), foam formation (FO)	Indicates bacterial activity in culturing fluids of tester; reductive fermentative activity leads to generation of gases including carbon dioxide, hydrogen, methane, and nitrogen
3	Brown clouding (BC), brown ring (BR), basal gel (BG), red clouding (RC), green clouding (GC)	Iron accumulated within biocolloids created by some FOB; FOB, RIB, and SIB are all capable of becoming significant when ferric-rich slime ring forms; some enteric bacteria (Enterobacter and Klebsiella species) generate a dense ferric-rich gel in the tester; some enteric bacteria (including Serratia species) generate red color in clouded growth; starts as weak CL reaction that slowly turns light green; may turn darker and cloudier over time; occurs when Pseudomonas species are very active
4	Black liquid (BL)	Walls and base of tester are coated with black materials (possibly dominated by iron carbonates), commonly when both enteric and pseudomonad bacteria are active

Note: sequence refers to the order in which reactions are likely to occur over the test duration; 1 is likely to occur first; 4 is the final reaction. This order requires that microorganisms are present and active enough to cause the specific reactions.

BC	Iron-related bacteria active
BG	Dense slime-forming slime forming bacteria located
BR	Very aerobic slime-forming iron-related bacteria observed
GC	Heterotrophic pseudomonad-type bacteria appear dominant
RC	Enteric bacteria very common
BL	Dominant mixture of pseudomonad and enteric bacterial species; possible health risk; conduct a coliform test

The term "iron-related bacteria" implies strongly that all the microbes cultured using IRB BART tester are likely to be involved in excessive manipulation of iron beyond basic metabolic needs. For microorganisms to be cultured and recognized by IRB BART, a number of constraints reduce the probability of non-IRB growing within the culturing fluids (Table 9.2). The dried nutrient pellet in the base of the tester dispenses chemicals that restrict the ability of many other bacteria to grow. The use of a crystallized pellet to deliver selective nutrients gradually increases the chemical concentrations as the chemicals diffuse upward through the culturing fluids. This forms a moving diffusion front that further limits bacterial activity to the IRB.

IRB BART generates, when charged with a suitable sample, a sufficient diversity of environments that encourages the development of observable activities of IRB in the sample tested. Based on experience to date, IRB BART appears superior to other

TABLE 9.2
Chemical Factors for Defining Iron-Related Bacteria

Ingredient	Restricting Function
Ferric iron	Ferric iron provided in sufficiently high concentrations to be inhibitory to bacteria with low tolerance; ferric form commonly reduced to ferrous form by IRB that allow iron cycle to occur
Citrate	Most IRB utilize citrate as carbon source; citrate acts as functional restrictor for heterotrophic bacteria that cannot use it as a carbon source
Nitrate	Alternate electron acceptor to oxygen; many IRB "respire" using nitrate; it provides improved selectivity for growth of IRB
Dipotassium hydrogen phosphate	IRB store phosphate primarily as polyphosphates; high phosphate concentration restricts the range of growing bacteria to those that are phosphate-tolerant

field-applicable testing systems based on the broad scope of recoverable IRB. Note that these claims are subject to the following limitations:

1. The limit of detection for the IRB in a given water sample is 67 cells per liter.
2. Any water sample taken for IRB BART analysis must be collected following the protocols established for collecting water samples for microbiological analysis. Transportation and storage of the sample should follow the standard guidelines for sample handling prior to the initiation of microbiological examination: aseptic handling, the use of sterile sample containers, and minimizing the sample storage time to less than 4 hours at room temperature or 24 hours under refrigeration.
3. IRB BART can be used for both field- and laboratory-based investigations and generates similar time lapse and reaction data when a sample is split and incubated under similar conditions in field and laboratory settings.
4. While the IRB BART operates at ambient (room) temperature, the testers may be used at incubation temperatures (+1 to +30°C) under exceptional circumstances. Incubation above 30°C is not recommended because of possible distortions.

9.5   BLACK CAP: SULFATE-REDUCING BACTERIA (SRB BART)

Sulfate-reducing bacteria reduce sulfate by producing gaseous hydrogen sulfide that reacts with iron to generate black iron sulphide. These bacteria also generate black slimes and start corrosive processes. SRB are thought to cause corrosion of fabricated iron equipment and structures, steel pipelines, storage tanks, and even concrete. A brief definition below applies to major SRB that use different sulfur-based substrates. Some SRB produce spores. The definition of SRB is:

All bacteria that reduce sulfate or sulfur to hydrogen sulphide; they usually are active under oxygen-free (i.e., reductive, anaerobic) conditions and use fatty acids (particularly acetate) as the main source of organic carbon.

SRB include the dominant genus Desulfovibrio. The bacteria can be found in two locations. First, they are found at the redox front between oxidative and reductive conditions, usually with ORP values between +10 and –50 mv. Extensive heterotrophic bacterial activity reveals a tendency to live deeper within the biofilms created by the aerobic bacteria. Second, some SRB grow in much more reductive environments (ORP values from –20 to –150 mv. Desulfovibrio is the most common genus associated with corrosive processes.

To detect sulfate-reducing bacteria (SRB) SRB BART uses a broad spectrum Postgate medium that can detect the full range of common hydrogen sulfide generating bacteria including SRB. Functionally, this is achieved by the establishment of an oxidation–reduction gradient when the BART ball restricts oxygen entry from the headspace downward into the sample contained in culturing fluids. The lower regions of the sample in the tester become reductive and the upper regions remain oxidative. These conditions generate one of two SRB reaction patterns (Table 9.3). Before adoption of the test method, comparisons with standard tests revealed the SRB tester exhibited greater sensitivity, was easier to use, and provided more reliable data than the industry "nail in the bottle") standard. Two major advantages revealed were: (1) no dilution of sample is required; the time lapse generated gives a direct link to the population: and (2) no need to adjust salinity; the test is performed in the original sample at ambient salinity.

The tester contains a Postgate medium that selectively encourages significant growth of SRB. Table 9.4 lists chemicals employed in the tester. SRB BART involves only two reactions of significance. In the black top (BT) reaction, the generation of 1 to 2 mm small jet black specks on the lower part of the BART ball indicates a positive. For the black base (BB) reaction, a jet black region formed in the cone at the base of the tester (see Table 9.3) indicates a positive. To view this, lift the tester and examine the cone-shaped base to find blackening. Sometimes small regions near the center blacken fairly quickly; they should not be considered positive results. Only after the blackening spreads over half the area of the base cone should results be considered positive.

BB deposits grow gradually and cover the whole base cone; they may even climb 2 to 4 mm up the side walls of the tester. A third reaction (no longer recognized) was designated black all (BA). It produced concurrent black sulfide reactions at the

---

TABLE 9.3

Sulfate-Reducing Bacteria Reaction Patterns and Diagnostic Comments

Reaction	Description	Comments
BT	Black specks or dark grey film on lower side of ball	SRB grow in association within bacterial consortia dominated by aerobic bacteria that function under oxidative conditions
BB	Radial blackening of cone in base of tester continues until whole base is black	SRB grow covertly under reductive conditions without growing in association with other bacterial species

TABLE 9.4
Selective Chemical Ingredients Used in SRB BART

Chemical	Comment on Selection
Sulfate	Reduced by SRB to hydrogen sulfide
Ferrous iron	H₂S generated by SRB reacts with ferrous iron to produce black sulfide, indicating positive SRB BART
Acetate	Major carbon substrate used by SRB for growth and metabolism; MPB may compete with SRB, causing significant gas production; tester conditions favor SRB over MPB
Anoxic blocker	Restricts diffusion of oxygen around ball; allows faster setup of reductive conditions; encourages BT reaction

top and the bottom of the tester. This reaction may be noted if the tester is visually examined daily, but the reaction should be designated BB. If the VBR72 system is used, the precise TL to the BB and BT reactions is determinable. Usually BT or a BB reaction precedes a BA. Interpretation is based upon the first reaction. BA is no longer an acceptable reaction. If a BA reaction occur without prior observation of a BT or BB, the reaction defaults to a BB. BT and BB reactions signify the presence of different communities of SRB. Both reactions involve Desulfovibrio and Desulfotomaculum species. The latter SRB produce spore that can survive higher temperatures than cells.

9.6 LIME GREEN CAP: SLIME-FORMING BACTERIA (SLYM BART)

Slime-forming bacteria (SLYM) produce copious amounts of slime (glycocalyx) without necessarily accumulating iron. These slime-like growths are dominated by greys, yellows, oranges, and occasionally reds generated by the bacterial pigments. SLYM can also function near different ORP conditions generated by biomass and usually produce the thickest slime formations under aerobic (oxidative) conditions. The formations can grow around a BART ball as slime rings.

Growth can also appear as cloudy (fluffy or tight plate-like) structures in culturing fluids or as gel-like growths that may be localized or appear throughout a sample. Gel-like slime growths in a tester usually form from the bottom upward. One common check for these types of growth is to tilt the tester gently and see whether the cloudier gel- like growths retain their structures and move with the tube. Most slime-forming bacteria produce copious amounts of slime that contributes to symptoms such as plugging, efficiency losses in heat exchangers, and clouding, bad taste, and odor in water. SLYM is one of the most sensitive BART analyses. A positive commonly involves a cloudy reaction in the tester, often with thick gel-like rings around the ball. A negative test remains clear. Most bacteria produce some form of slime-like growth. The slime is actually formed by extracellular polymeric substances (EPS), long, thread-like, stringy molecules that bind to water. The EPS coat the cells and form slime masses in which large volumes of water are clustered and bound and hold the cells tightly. About 95 to 99% of slime volume is water. Some bacteria produce EPS that remain tightly bound to individual cells. These are called capsules. Other bacteria generate so much EPS that it envelops masses of cells within an individual slime.

Slime appears to have a protective role for the bacteria and a conservational role. If environmental conditions are harsh (e.g., shortage of nutrients), the slime layers thicken. The slime acts as a protectant to the resident bacteria and also acts as a biosponge by accumulating many chemicals that could form nutrient bases or be toxic to the cells. EPS may be produced by enzymatic activity (dextran sucrase or levan sucrase) on carbohydrates. EPS may be synthesized within the bacterial cells and released to form an enveloping slime. Reaction patterns for the SLYM BART are:

Dense slime (DS) — This reaction may not be obvious and may require the observer to gently rotate the SLYM BART. Slimy deposits, possibly in twisting form, may swirl upward. The swirl can reach 40 mm up into the culturing fluids or it may rise as a group of globular gel-like masses that settle fairly quickly. After the swirl settles, the liquid may become clear again. If that occurs, care should be taken to confirm that the artifact is biological (ill-defined edge, mucoid, globular) rather than chemical (defined edge, crystalline, often white or translucent). Dense slime growths are normally beige, white, or yellowish-orange. This reaction is defined as the formation by DS bacteria of copious EPS, with facultative anaerobes dominating.

Cloudy plate (CP) — This layering appears when populations of aerobic bacteria are present. The initial growth may be at the redox front that forms above the yellowish-brown diffusion front. Growth usually takes the form of lateral or puffy clouding, usually grey. The lateral clouds may be disk-like (plates) and relatively thin (1 to 2 mm). If the observer tips the BART slightly, the clouds or plates often move to maintain positions within the tube. The edges of the plates are distinct. The edges of the puffy layer forms are indistinct. The formations are usually observed 15 to 30 mm beneath the fill line. Cloudy formations may extend to cause overall cloudiness of the liquid medium (CL). The plates appear to divide (multiple plating) before coalescing into a cloudy liquid medium.

Slime ring (SR) — These rings are usually 2 to 5 mm in width and form on the upper side of a BART ball. Appearance is mucoid and a ring may be white, beige, yellow, orange, or violet. Color on the upper edge becomes more intense over time.

Cloudy growth (CL) — The reaction solution is very cloudy and a poorly defined slime growth will appear around the ball. A glowing area may appear in the top 18 mm of the liquid medium. The glow arises from ultraviolet fluorescent pigments generated by some Pseudomonas species. The common pigments are greenish yellow (see GY below) and pale blue (see PB below). Note that this glowing may not be readily observable unless ultraviolet light is used. Glowing revealed by ultraviolet light means a probability of potentially pathogenic Pseudomonas species. Confirmatory testing is recommended;

Blackened liquid (BL) — This result is commonly a secondary or tertiary reaction, not an initial reaction. The liquid is clear, often colorless, and surrounded by large blackened zones in the basal cone and up the walls of the test vial. The BL often parallels the BL reaction of IRB when two BART testers are used to test one water sample.

Thread-like (TH) — Slime may generate threads or strands that form web-like patterns in liquid media. The threads may interconnect from the BART ball to the floor of the tester.

Greenish-yellow (GY) — This type of glow is easily seen under ultraviolet light. It usually follows the CL (cloudy) reaction and forms around the BART ball. It appears after 2 to 4 days and may extend halfway down the tester. The GY reaction indicates the possible presence of *Pseudomonas fluorescens*.

Pale blue (PB) — This glow visible under ultraviolet light usually follows the CL (cloudy) reaction. The distinctive pale blue color immediately appears around the BART ball and usually remains commonly for less than 2 days. The PB reaction indicates the possible presence of *Pseudomonas aeruginosa*.

9.7   DARK BLUE CAP: HETEROTROPHIC BACTERIA (HAB BART)

Heterotrophic bacteria are defined by their ability to exploit and biodegrade organic materials as energy sources: nitrogen and phosphorus. The HAB group are almost universally present in the alpha two consorms. Various species have specific requirements, but as a community HAB possess the resources to utilize carbohydrates, lipids, proteins, and even hydrocarbons such as methane (in alpha five bacterial consorms). For this reason, they often dominate when spills of solvents or hydrocarbons occur and play major roles in degrading toxic organic materials. The culture used for HAB BART is a crystallized proprietary LCP medium rich in diverse carbohydrates and proteins.

While developing HAB BART, it was found that the UP reaction (methylene blue [MB] decolorizes or bleaches from the bottom up) tends to occur when heterotrophic bacteria growing aerobically dominate. Such aerobic events are likely to occur immediately downstream of organic pollution involving oxidative microbial activity. DO reactions (MB decolorizes or bleaches from the top down) by contrast appear in waters dominated by facultatively and strictly anaerobic heterotrophic bacteria that are active under more reductive conditions. A sample is most likely to be from the reductive side of an oxidation–reduction interface. Therefore DO reactions imply that the biofouling is from a reductive zone (where oxygen is absent). An UP reaction implies major oxidative activity at the site because of oxygen in the water.

One unique feature of HAB BART is that in addition to a specific enriched mineral and nutrient medium, MB is also added as an ORP indicator. This addition allows TL monitoring of respiration the initiation of "bleaching" (MB changes from blue to clear or yellow). While free oxygen remains in the culturing fluids, the MB in the liquid medium remains blue. As soon as all the oxygen is consumed by bacterial respiratory activity faster than oxygen can diffuse into the site, the MB shifts from its observable blue to yellow or colorless. In summary, when the liquid medium turns from blue to a yellow or colorless form in HAB BART, the heterotrophic aerobic bacteria are sufficiently active to "respire off" all the oxygen from that zone in the tester. The TL to the bleaching action correlates with the concentration (population) of heterotrophic bacteria in the sample.

Dried methylene blue is present in the cap of each HAB BART tester dissolves into the liquid sample when the tester is inverted for 30 seconds followed by three wrist action inversions. During this process, the BART ball moves up and down the test vial a total of six times. This allows the head space air time to saturate the turbulent culturing fluid samples with oxygen and achieve saturation.

MB is a basic dye that can readily bind to negatively charged bacterial cells. Traditionally, it is used to stain microbial cells. The important property of MB is the change from the original blue in the oxidized state to clear or yellow when reduced (oxygen absent). When MB is added to a liquid medium at respirable organic concentration, electrons are transferred to the dye molecule, causing it to reduce, blue (oxidation) color disappears. The rate at which this happens depends on the rate of microbiological respiratory activity. This feature allows the operator to determine the size of an active heterotrophic bacterial population.

With the UP reaction, the MB solution decolorizes from the bottom up. The bleached zone underneath will appear clear. A clouded yellow zone relates to a light to medium yellow color already in the culture medium. Rarely does the bleaching extend upward beyond the equator of the BART ball. A blue ring normally remains around the ball 1 to 5 mm below the surface after the completion of the test. With the DO reaction, the bleaching usually forms just below the BART ball in a zone 32 to 42 mm above the base of the tester. Unlike the UP reaction, the DO forms as a series of swirling, unstable decolorized regions in the medium that then stabilizes into a clear colorless lateral zone with a front that moves down the liquid column.

MB impacts heterotrophic bacterial activity. In a series of trials of the medium with and without MB, consistent differences were noted. The controls in three replicated trials had time lapses of 1.62 ± 0.51 days while the MB (in the HAB BART) caused a delay and yielded a TL of 2.47 ± 0.77 days. This difference of 0.85 ± 0.26 days is attributable to the inhibitory effect of MB on the active bacterial population.

For the control, the method for detecting activity was visual (formation of turbidity). In the HAB BART, the VBR72 system directly detected changes in sorption as the MB reduced. Since turbidity almost always follows the development of reductive conditions (MB reduction in HAB BART), the probability is that MB exerts an inhibitory effect on the heterotrophic bacteria that causes prolongation of the TL. This effect appears transitory since in all cases observed there was subsequent activity with the HAB BART. It is therefore claimed that MB acts as a transitory inhibitor of heterotrophic bacterial activity. This impact consistently increased TL by an average of 52%.

The lowest number of detectable cells (1 cell per 15 ml or 67 cells per liter) would be needed to create a positive TL that would relate to the number of HAB cells that became active under incubating tester conditions during the test period. This forms the theoretical basis of claims that the length of the TL (in seconds) is inversely related to the number of active cells in the sample tested.

9.8 GREY CAP: DENITRIFYING BACTERIA (DN BART)

Denitrification (DN) is the reduction of soil nitrate or nitrite to nitrogen gas. This activity is extremely important in the environmental, agricultural, and geochemical fields. The reason is that essentially all atmospheric nitrogen (N_2) has been derived from bacterial DN. Thus DN is a vital stage in the nitrogen cycle on the surface and immediate subsurface biospheres.

In a specific cycle, nitrogen from the atmosphere is fixed, cycles through the biomass to ammonium which is then oxidized to nitrate by nitrification (see N BART) and reduced back to gas by DN. The final step (nitrate to nitrogen gas) is controlled by denitrifying bacteria. The denitrifying bacteria are important indicators of the decomposition of organic nitrogenous wastes. In waters, an aggressive population of denitrifiers is an indication of significant nitrate in the water. When DN BART detects denitrifying bacteria in a sample, the reaction occurs as a ring of foam (FO) around the BART ball. This FO reaction generates dinitrogen gas that foams around the BART ball before dissipating (usually after 24 hours).

A common use for aggressive denitrifying bacteria in waters is signaling the degradation of nitrogen-rich sewage and septic wastewater through the post-treatment oxidative phase. Aggressive denitrifiers in water can indicate the potential for the groundwater to have been polluted by oxidized nitrogen-rich organics from such sources as compromised septic tanks, sewage systems, and industrial and hazardous waste sites. Where highly aggressive conditions are found, the recommendation is to subject the water to further evaluation as a hygiene risk through confirmation of the presence of coliform bacteria and nitrates. In soils, an aggressive denitrifying bacterial population may indicate that the denitrification phase of the soil nitrogen cycle is functioning. One would expect a greater potential of denitrifying bacteria in groundwater under the direct influence of surface waters where the water moves across the redox front and contains significant concentrations of inorganic and/or organic nitrogen.

DN reduce nitrates to nitrogen under anaerobic conditions. This is significant because nitrates in water represent a serious health concern, particularly for babies. DN decrease this risk when the environment is reductive. To test for DN, a gas forms FO from nitrogen gas bubbles that usually collect around the ball within 3 days. The appearance of FO by the end of day 3 indicates an aggressive population of DN. The absence of foam, regardless of clouding of the water, indicates that the test did not detect denitrifying bacteria. This test is applicable to any samples likely to contain potentially septic or organic contamination. For example, denitrifiers in water indicate a potential health risk from septic wastes or nitrates in the water. The test detects bacteria that can reduce nitrate (NO_3) to nitrogen gas (N_2). Gas (through the formation of FO) is generated when the nitrate has been completely denitrified.

9.9   WHITE CAP: NITRIFYING BACTERIA (N BART)

Nitrifying bacteria are important indicators for the recycling of organic nitrogenous materials from ammonium (the end point for reductive decomposition of proteins) to the oxidative production of nitrates. In the environment, aggressive populations of nitrifiers indicates the potential for the generation of significant amounts of nitrate in aerobic (oxygen-rich) waters. Nitrates in water cause concern because of the potential health risk, particularly to infants who have no tolerance to nitrates. In soils, nitrification is a very significant phase of the recycling of nitrogen through soil.

Nitrate is a highly mobile ion in soil and moves (diffuses) relatively quickly while ammonium remains relatively locked in the reductive regions of soil.

In some agronomic practices, nitrification inhibitors are used to reduce losses of ammonium to nitrate and improve crop uptake. A common use is to counteract nitrifying bacteria with an inhibitor applied to the environment to halt the oxidization of ammonium to nitrate. This oxidization signals late stages in the aerobic degradation of nitrogen-rich organic materials by the oxidation of ammonium to nitrate, often after the reductive degradation of proteins with the production of ammonium. Active nitrifying bacteria indicate nitrogen-rich organics under reductive conditions from such sources as compromised septic tanks, sewage systems, and industrial and hazardous wastes. Ammonium is a product of such reductive proteolysis and is oxidized to nitrate through nitrification.

Nitrification and denitrification are essentially parallel but contradictory processes that function in reverse sequences (cycles). If high activity is determined, samples should be subjected to further evaluation as hygiene risks through a subsequent investigation for nitrates. In soils, an active nitrifying bacterial population may indicate that the nitrification phase of the soil nitrogen cycle is functioning. Nitrification is fundamentally an aerobic process in which the ammonium is oxidatively converted to nitrate via nitrite, The reverse denitrification process is anaerobic (reductive), Nitrite can also be produced by the denitrification of nitrate and can be oxidized back to nitrate. The steps of nitrification are:

$$NH_4 \,(\text{ammonium}) \rightarrow NO_2 \,(\text{nitrite}) \rightarrow NO_3 \,(\text{nitrate})$$

Step 1 is the transformation of ammonium to nitrite; in Step 2, nitrite is oxidized to nitrate. Nitrification serves as the major route by which ammonium is oxidized aerobically to nitrate. Nitrifying bacteria are divided into groups according to their functions:

Group	Step	Function	Genus
I	1	Nitrosofiers	Nitrosomonas
II	2	Nitrifiers	Nitrobacter

The contradictory relationship between nitrifying and denitrifying bacteria presents a problem in testing natural samples both groups can produce *and* utilize nitrate. In a test to determine nitrifying bacteria in natural samples, the terminal product (nitrate) may not be recoverable with certainty because of the intrinsic activities of the denitrifying bacteria likely to be present and active in the sample. Because of the dual tasks of these bacteria, N BART is restricted to detecting the nitrosofiers that generate nitrite. The nitrite is oxidized to nitrate by the nitrifiers, only to reappear when reduced back to nitrite by any intrinsic denitrification in the sample.

N BART follows an unusual protocol in that the nitrifying bacteria are detected by the generation of nitrite in the test vial after a standard incubation of 5 days. Nitrification involves the oxidation of ammonium to nitrate via nitrite. Unfortunately, in natural samples, denitrifying bacteria in a sample reduce the nitrate back to nitrite under reductive conditions. When denitrification is complete, the nitrite is reduced further to nitrogen gas (under anaerobic conditions). For this reason, N BART testers are unconventionally laid on their sides. The inner vial of the tester contains three BART balls to provide a moistened, highly aerobic upper surface where nitrification will most likely to occur. After 5 days of incubation, the diagnostic reagent is administered in the reaction cap and detects nitrite specifically by a pink-red reaction in 3 hours. Interpretation is based on the amount and location of the pink-red coloration (see below), This test is different from the other BART techniques in that a chemical reagent is added to detect the product (nitrite) after a standard incubation period. The typical reactions are:

Partially pink on balls only (PP) — A clear solution with a pink reaction may be generated on the BART balls. This indicates nitrification; the nitrite detected is in the biofilm on the balls. The pink-red color covers roughly half the ball; the solution is clear or yellow.

Red deposits, pink solution (RP) — The reaction produces a light pink solution with red deposits spread over the three BART balls. Nitrite is present in solution and in the biofilms on the BART balls. All three balls are reddened and the solution may be pale pink.

Dark red deposits and solution (DR) — This reaction changes the solution to turn dark red with heavy red deposits on the BART balls. High concentrations of nitrite indicate an aggressive level of nitrification during the test period.

9.10   PURPLE CAP: ACID-PRODUCING BACTERIA (APB BART)

Acid-producing bacteria (APB) include heterotrophic bacteria that share a common ability to reductively produce organic acidic products (mainly fatty acids). APB cause the pH to drop significantly from neutral to acidic; terminal pH levels range from 3.5 to 5.5. These mildly acidic conditions are sufficient to be significantly corrosive to metallic alloys and concrete structures. Because these acid-producing activities occur in the absence of oxygen, APB are likely to be significant partners in corrosion with SRB, particularly at oil and gas facilities. The management and control of corrosion require assessment of the aggressivity of both APB and SRB.

Microbiologically influenced corrosion (MIC) was considered to be dominated by SRB because of their ability to trigger electrolytic corrosion of metals, primarily under highly reductive conditions in the presence of adequate sulfates and organics. SRB generate hydrogen sulfide (H_2S) as a metabolic end product and the H_2S triggered the electrolytic corrosive processes. It was later observed that some MIC was acidulolytic, caused by bacteria able to generate acidic products generally under highly organic and reductive conditions. Little is known of the complex natural activities of bacterial consorms with respect to the intelligent manipulation of electrolytic forces as a part of their natural metabolic functions.

APB are significant contributors to corrosive processes because of gradual dissolution of metals and concretes under the acidic conditions they create at the metal–biomass interface. APB communities are active under reductive conditions within biofilms, slimes, encrustations, ochres, nodules, and tubercles. Their activity may create lateral erosions of metal surfaces most clearly seen if the metal surface is examined under reflective light. If APB are present, a metal surface will exhibit an irregular pattern of shallow depressions. The most effective examination for APB requires sampling a slime, concretion, encrustation, nodule, or tubercle at the metal–biomass interface (between the growth and metal surface). The result of the activity differs from the electrolytic corrosion caused by SRB that tends to cause deeper pitting of the metal and a greater risk of perforation. In simple terms, APB can generally impact metal surfaces; SRB cause focused pitting, cavitations, and perforation.

Reductive conditions mean that APB always generate organic acids fermentatively, but the activities do not usually drop the pH below 3.0. Other bacteria generate inorganic acids (e.g., sulfuric) as a result of the oxidation of sulfides and sulfur. These bacteria are associated with acid mine drainage and cause downstream waters to become very acidic through the oxidation of sulfides and/or sulfur to sulfuric acid. The major genus associated with these activities is Thiobacillus.

To conduct APB BART, the tester is placed upright at room temperature and observed daily by an operator of monitored via a VBR72. If APB is present in a sample and becomes active, the pH will drop in the more reductive regions of the tester and the culturing fluid color will turn from purple to dirty yellow. This color change indicates a pH drop to the 3.5 to 3.9 range. This acidic condition will progress over time until most of the liquid in the tester turns yellow. A TL occurs when the tester liquid is yellow (acidic). Occasionally, a buffering action will also be seen in the tester, causing the liquid to return to the purple color. This buffering is brought about by oxidative (aerobic) activity around the BART ball, extending downward into the tester. The buffering does not mean that APB were not present; it means that the consorm included many heterotrophic aerobic bacteria that buffered the generated acidity.

9.11 YELLOW CAP: FLUORESCENT PSEUDOMONAD BACTERIA (FLOR BART)

FLOR BART was developed to detect heterotrophic bacteria species belonging to genus Pseudomonas. Some species have been linked to increased health risk; others dominate bioremediation through their ability to degrade organic pollutants efficiently under oxidative (aerobic) conditions. When water samples are tested using FLOR BART, the main focus is the production of colors that fluoresce in commercially available ultraviolet light. The two colors detected are: (1) pale blue glow (PB); and (2) greenish-yellow glow (GY).

These two reactions are generated by particular species of Pseudomonas. In the PB reaction, the glow is seen mostly around the BART ball, extending 2 to 6 mm beneath the ball. The color reaction occurs 2 to 4 days after the start of the test and then fades. At its brightest, the PB can be seen under regular room light. The glow of the GY reaction this will be seen mostly around the BART ball, extending 4 to 10 mm beneath the ball. This color reaction occurs 3 to 10 days after the start of the test and can last 2 weeks before fading.

A PB reaction indicates *Pseudomonas aeruginosa* are likely to be present. This species presents a potential health risk. *Pseudomonas aeruginosa* in water may cause infections in humans—pneumonia, eye and ear infections, and even skin and lung infections. When PB is observed, confirmatory testing should be performed by a certified analytical microbiology laboratory. A GY reaction indicates *Pseudomonas fluorescens* are likely to be present. This species frequently dominates waters in which significant aerobic degradation of specific organics takes place.

9.12 COMPARISON OF BART AND OTHER BACTERIOLOGICAL ENUMERATION METHODS

BART methods provide relatively simple ways to detect specific bacterial communities involved in consormial activities and determine how active the communities are. Inverse correlations of TL and population size make the BART system superior for quantitative monitoring. TL serves as the primary indicator of the level of microbial activity in a sample. Comparisons of BART and other microbiological test methods are discussed below.

9.12.1 COMPARISON OF BART, HPC, AND ATP

There has been some suggestion that the BART technologies cannot be compared with the ATP (adenosine triphosphate) and HPC (heterotrophic plate count) methods. Table 9.5 presents a direct comparison of the relative advantages and disadvantages of all three methodologies.

All living cells produce adenosine triphosphate (ATP) as the main mechanism to store energy. A sample containing many active microbial cells also contains ATP.

TABLE 9.5
Comparison of HAB BART, HPC, and ATP

Feature	HAB BART	ATP	HPC
Target	All active heterotrophic bacteria in sample	All active prokaryotic or all cell types active in sample	All active heterotrophic bacteria able to grow on agar
Initial introduction	1991; Canadian ETV in 2003	1980s with EPA acceptance	1930s with modifications through 1980s
Differentiates aerobes from anaerobes	Yes	No	No
Field applicability	Yes	Can detect activities of bacteria in sample	No; must be set up in certified laboratory
Duration of testing	Up to 5 days	Rapid in laboratory with correct methodologies	Requires skilled operator; 1 to 4 days
Operator skill level	One day training	Certified technologist	Certified microbiological operator
Sample handling	Test performed within 4 hours at site	Sample shipped to certified laboratory	Sample must be shipped to certified laboratory

Testing for ATP in a sample has become the "gold standard" for counting active cells. BART methods use a similar approach but measure activity by the TL of a recognized reaction or activity. BART uses a selective culture medium rather than the blanket approach of measuring ATP concentration. A shorter TL indicates more bacterial activity and higher ATP levels.

9.12.2 Environmental Technology Verification (Canada)

BART technologies have been subjected to evaluation in Canada. Three systems have achieved Environmental Technology Verification (HAB, SRB, and IRB). The three systems have been in routine use since 1991 and have been adopted by the U.S. Army Corp of Engineers to evaluate the effectiveness of treatments of all kinds of water wells. Further information is available at www.dbi.ca. Note that:

1. Independent comparisons of TL and APT data were conducted by the Universities of Western Ontario and Saskatchewan and correlations were obtained.
2. Further information about BART can be found in *Microbiology of Well Biofouling* (Lewis Publishers/CRC Press, 1999); *Simplified Atlas for the Identification of Bacteria* (Lewis Publishers/CRC Press, 2000); and *Practical Manual for Groundwater Microbiology, Second Edition* (CRC Press, 2008).
3. Comparison of the BART systems with the various other microbiological test methods is detailed on the Web site.
4. Over time, BART methods have proven reliable and sensitive to active bacterial populations. This feature caused problems when BART procedures showed that treatments by companies that made overly optimistic claims for their products did not produce sustainable impacts on bacterial populations.

9.12.3 Comparison of BART and Other Microbiological Tests

BART is used directly on samples and can accommodate solid or semi-solid samples. No dilution is required if a liquid sample has low turbidity. The tester is incubated at room temperature ($22 \pm 2°C$) although $28 \pm 1°C$ is faster and generates greater precision. BART accommodates 15 ml of liquid sample for the test and measures activities or reactions caused by any microorganisms that grow in the sample. Solid and semisolid samples (e.g., soils and muds), 0.1 to 0.5 g, are diluted into a sterile diluent. In testing of liquids where no dilution is required, impact on the microorganisms in a sample is minimized and improves potential for effective recovery. The tester device contains a variety of environments to increase the potential for activities and reactions to occur. Comparisons of other technology are described below:

Agar spreadplates — Agar is used to make a jelly-like base upon which bacteria can grow to form colonies (visible piles of cells) that are easy to count and (with the proper agar) can be used to identify the types of bacteria present. That is why counts on agar-based media are expressed as colony-forming units (CFU). A disadvantage of agar plates is that the microorganisms must grow in a form recognizable as a distinctive colony. Despite its convenience, this method does not detect the many types of bacteria that cannot grow on agar, cannot tolerate high levels of oxygen, or cannot extract water effectively from agar. BART methods allow many more types of microorganism to grow, often much faster because of the greater variety of environments that the tester presents to the organisms in samples. BART is more sensitive to a wider variety of microorganisms than agar-based media and delays before growth occurs are shorter.

Pure bacterial cultures and standard plate count agar techniques were used to compare TLs generated by BART with colony counts on agar media. Under these controlled conditions, the predicted active cells per milliliter (pac/ml) correlated to the colony forming units per milliliter (cfu/ml) and the populations were considered comparable.

Agar dip paddles — These devices consist of a relatively thin agar film deposited on a plastic paddle that is immersed in a water sample to inoculate the agar surface with microorganisms. Microorganisms must become attached to the agar surface and grow to form visible colonies when the dip paddle is incubated. The agar forms a very thin film over the plastic and can dry out quickly, increasing the concentrations of chemicals in the agar and reducing the range of microorganisms that can form colonies. BART uses 1.5 ml of sample and presents better potential for cultivating bacteria efficiently. The result is more precise data without concerns about dried agar that may produce inaccurate results.

Quick (color change) tests — A number of fast microbiological tests claim to detect bacteria semiquantitatively (e.g., a lot, a few, none). Such tests usually involve filtering or contacting a sample to trap the microbial cells, then staining with a colored reagent that reacts to the cells. The more cells present on the filter, the more intense the color. The problem is that filtration concentrates all the organisms (live and dead) onto the filter. The reagent may become reactive to the dead organic matter and signal a falsely high color reaction that does not correctly indicate the number of active cells present. While these tests are fast, they lack the precision of BART, particularly when the VBR72 is incorporated. Some BART methods such as HAB BART can react as quickly as 3,000 to 5,000 seconds when very large active populations of bacteria are present (as in primary influent from a sanitary waste water treatment plant) and results are reproducible. BART is quick, particularly when a large and active bacterial population is present in a sample.

Immunoassay — A new science called immunology was developed in the 1920s based on the finding that all microorganisms developed unique chemical signals. Some signals were specific to disease-producing microorganisms and caused specific responses that enabled the body to become resistant (immune) to pathogens. Developments in biochemical science allow many microorganisms to be identified precisely by unique chemical signals carried by the cells of specific species. Some of these chemicals are called antigens. An infected body produces antibodies that neutralize them.

By applying these antibodies that in essence wear color-coded tags, we can detect specific species of bacteria. These tests are more effective in a laboratory setting with highly trained technologists, but field tests are available. An immunoassay for

SRB may involve six or more steps, precise timing, and various devices to achieve a result. The highly precise technique required by the field test opens the door for a variety of errors leading to false positives or negatives. BART examines samples of whole communities of bacteria and seeks to quantify the activity levels (population size) and reactions that occur—qualitative determination of the types of bacteria present. Immunoassay tests are designed to detect specific species of bacteria in a sample. Most samples contain active communities that can include as many as 8 to 16 species in each community. BART provides more information with a simpler method. Immunoassays are complicated and expensive; BART is simple, economical, and can detect nuisance bacteria that cause problems.

10 Introduction to Grid-Formatted Bacteriological Atlas

10.1 FOCAL POINT LOCATIONS FOR BACTERIAL CONSORMS

This atlas utilizes a bacteriological positioning system (BPS) based on oxidation–reduction potential (ORP) expressed in millivolts (fmv). ORP forms the x axis (latitude) and water viscosity as the log factor of centipoise (fcP) forms the y axis (longitude). Each bacterial consorm is defined by its focal point location (FPL) expressed as fmv:fcP). Its FPL locates it on the grid formatted bacteriological atlas. The x axis is subdivided into four major subgroups: (R = reductive; G = gradient; F = front (or interface); and O = oxidative. Each subgroup has six grid positions. This makes a total of 24 potential grid positions along the x axis representing ORP originally measured in millivolts and presented as fmv (fmv of 1 is most reductive; 24 is the most oxidative). The y axis defines water viscosity within which the consorm is active. Five units (0, 1, 2, 3, 4) are based on the log value of the centipoises (cP). Each unit is subdivided into six subunits. This creates a y axis with 30 units differentiated (5 groupings × 6 subunits). See Table 10.17 for the factorial relationships applied in calculating fmv and fcP.

In calculating the fcP value, zero indicates water in a dispersed form in the atmosphere. The fcP values of 1 through 4 indicate the viscosity of water bound in extracellular polymeric substances (EPS) or in chemical matrices such as clays. Thus the fcP reflects how tightly water is bound and such binding affects the ability of a bacterial consorm to function in an environment.

10.2 DIFFERENTIATION OF GRID ATLAS INTO SIX MAJOR CONSORMIAL GROUPS

Figure 10.1 represents the positions of all of the bacterial consorms recognized by designated focal point location (fmv: fcP). Each consorm has a three-letter designation code relating to its specific activity. Clustering of the consorms in the grid reflects their prime relationship to the FPL. Six major locations are defined for specific activity types; each has a distinctive FPL. The six recognized clusters were defined in Chapter 3.

Alpha Group	Consorm Description	FPL (Alpha Group, fmv:fcp)
1	Bionucleating dispersed	1, 22-04
2	Organic bioconcreting	2, 22-16
3	Inorganic bioconcreting	3, 13-21
4	Carbon reducing	4, 06-27
5	Carbon oxidizing	5, 13-07
6	Hyperbaric dispersed bionucleating	6, 01-03

Alpha two and alpha three are the largest of the six groups. Alpha two consorms occupy the surface biosphere and include many animal bacterial pathogens. Alpha three is associated with the subsurface biosphere and geochemical events. Most bacteriological findings relate to the alpha two bacteria because they produce such impacts on animals and indirect influences on plants. While the FPLs occupy specific grid positions, the six alpha clusters overlap to an extent.

Figure 10.1 shows the 24 × 30 grid plots incorporating in the x axis ORP conditions from very reductive (left) to very oxidative (right). The y axis represents the viscosity of the water from dispersed aerosol forms (0) to increasingly viscid water from 1 (free liquid) to 4 (tightly bound, highly viscid). Each grid position reflects a specific ORP and water viscosity. The six alpha groups are the middles of clusters of bacterial consorms that exhibit similar characteristics. A vertical points to the focal grid point of each major consormial group.

10.3 ALPHA ONE: BIONUCLEATING DISPERSED CONSORMS

The water in bionucleating dispersed clusters (1, 22-04) takes the form of particles dispersed in the air. Figure 10.2 shows the FPLs of the six consorms of the alpha one cluster. Three of the six consorms are dispersed into atmosphere as bionucleated particles: clouds (1, 22-03 CLD) and foam (1, 19-06 FOM) under very oxidative conditions. Foam also forms bubbles at the liquid water–interface. Lightning (1, 16-03 LGN) also occurs within the atmosphere but this is based on speculation. Negatively charged bionucleated particles may be impacted by friction created by winds and clouds, creating more reductive conditions that release electricity stored as biologically nucleated natural capacitors in clouds.

The three other bacterial consorms in alpha one inhabit water and are not dispersed in the atmosphere. Ice (1, 16-12 ICE) and snow (1, 19-12 SNO) are formed differently in liquid water; both are crystalline. Ice is crystallized water that collects at the water–air interface or may be a secondary product of the collapse of snow into water under pressure. Snow is composed of crystallizing water in which bacterial extracellular polymeric substances (EPS) act as surfactants to produce the shapes and forms of water crystals. Both ice and snow have fcP values of 12. The 12 indicates liquid water increasing in viscosity to the state of ice. Bacteria can be present in both ice and snow. In ice, they tend to remain active within voids of viscid liquid water. Such voids are common in ice subjected to organic-rich discharge waters from landfills and sewage treatment plants. Snow is formed by the directed crystallization of water into complex lateral patterns. It may or may not contain active bacteria but will always contain EPS generated as daughter products by bacterial consorms. Ice comets move

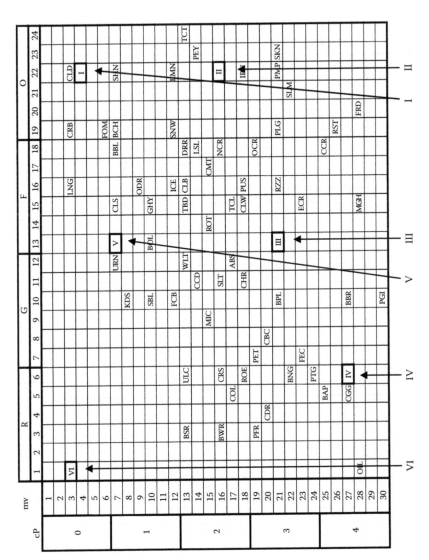

FIGURE 10.1 Grid atlas differentiation into six major consormial groups.

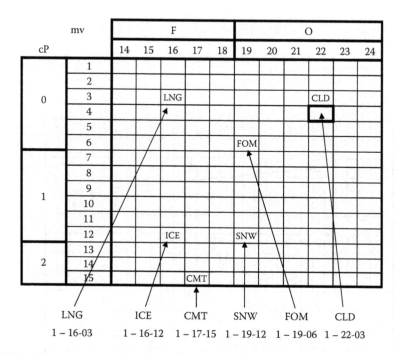

FIGURE 10.2   Grid atlas differentiation for 1, 22-04 alpha one consormial groups.

mv	R		G						F						O					
cP	6	7	8	9	10	11	12	13	14	15	16	17	18	19	20	21	22	23	24
7							URN		CLS					BCH			SHN		
8					KDS														
9																			
10						BOL													
11																			
12					FCB												LMN		
13							WLT		TBD	CLB		DRR							TCT
14						CCD												PEY	
15				MIC					ROT										
16						SLT						NCR					II		
17							ABS												
18						CHR				CLW	PUS								
19	ULC																		
20				CBC															
21										RZZ				PLG			PMP	SKN	
22																SLM			
23		FEC																	
24	ROE																		

FIGURE 10.3   Grid atlas differentiation for 1, 22-16 alpha two consormial clusters.

through space and likely contain active bacterial consorms designated provisionally as 1, 17-15 CMT. One speculation is that early life forms on Earth originated when ice comets inoculated surface environments when liquid water could freely settle there.

10.4 ALPHA TWO: ORGANIC BIONUCLEATING CONSORMS

All these consorms synthesize organic concretions—mixtures of cell structures and EPS. They accumulate few inorganic chemicals beyond their basic metabolic needs. These consorms are very diverse and adapt to many ORP and viscosity conditions to create very large clusters that center around 2, 22-16 as a focal point (see Figure 10.3). The three major clusters of consorms within alpha two are categorized by their preferences for oxidative (O), ORP front (F), or reductive (R) conditions. Figures 10.4 through 10.6 show the consorms within the three clusters (2O, 2F, and 2R).

Cluster 2O (Figure 10.4) includes nine bacterial consorms. Three are associated with free liquid water (fcP 1); two (BCH and SHN) relate to the interface between atmosphere and water; the third (LMN) occurs in EPS-bound water as suspended biocolloidal particles in the deep scattering layer or incorporated into light-emitting organelles on fish.

The thatch-forming consorms (TCT) thrive in very oxidative conditions (fmv 24); water is tightly bound into the biomass. These consorms actually "glue" plant materials together into an almost impermeable weave that deflects precipitating water, leaving the underpinning soil arid. The pink eye consorm affects the eyes. These consorms also exist in very oxidative conditions with some water viscosity.

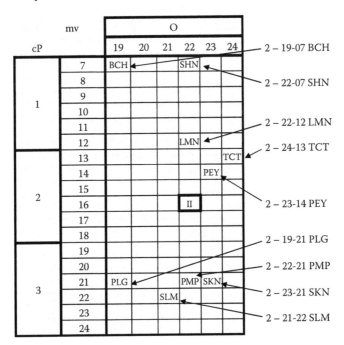

FIGURE 10.4 Grid positioning of alpha two consormial clustering 2O.

The remaining four cluster 2O consorms thrive in more viscid waters (fcP 3) that are still oxidative. Plugging consorms (2, 9-21 PLG) are very dynamic. They start in an oxidative environment (fmv 19) but gradually shift across the oxidation–reduction interface as the biomass controls the waters within voids and fractures. PLG consorms move during maturation from alpha group conditions (synthesising biomass) to alpha three (accumulating inorganic chemicals, primarily iron). The fmv levels of PLG consorms thus drop from 19 to as low as 7 (reductive), crossing the interface as the biomass shifts from alpha two to mainly alpha three activity. These consorms are dynamic and will shift to dominate the interface. The biomass eventually plugs the geological material impacted by the growth. PLG is an adaptable consorm that can adjust to many environments. The prime consequence is the loss of hydraulic transmissivity (bad for a producing water well, good if it stops a leak from a water line).

The remaining three consorms in 2O are tightly arranged, with fmv values of 21 to 23 and fcP values of 21 or 22. These consorms are most active in moderately viscid oxidative waters. PMP (2, 22-21) tends to form as an outgrowth from a living or dead surface. This outgrowth extends outward into the water (or gaseous) phase of the environment. Initial biomass growth is dome-like but can move to alpha three and begin bioaccumulating inorganics. SKN consorms (2, 23-22) form surfaces (skins) at water interfaces with solids or gases. SKN begins as an EPS-rich biomass that forms an integrated film over a surface. It can shift toward alpha three via the synthesis of carbonates and the accumulation of oxidized cations. Some SKN consorms develop significant electrical generation potential as a result of secondary maturation. SLM consorms (2, 21-22) generate excessive EPS-bound water within the biomass. The growth takes on slime-like characteristics and is usually most aggressive under oxidative conditions.

Ten consorms (cluster 2F) are involved in oxidation–reduction interface activities. They dominate; ORP ranges from 13 to 18 mv (Figure 10.5). Water viscosity varies from 7 to 21 cfP. Only two consorms are found in liquid-free water; the remainder (except RZZ) are in viscosity class 2 waters.

The 2F bacterial consorms aid in the generation of biomass at the oxidation–reduction interface. Water loses it clarity and becomes cloudy in a poorly defined manner as a result of 2, 15-07 CLS activities. The biomass is dominated by bacteria generating a diffusive EPS-bound water matrix. When held up to light, the water will be opaque but show no structure. A structure indicates turbidity (2, 15-13 TBD), visible by holding the water up to a light and looking for structures that resemble clouds in the biocolloids. Colloidal water (2, 15-18 CLW), when held up to light, should reveal viscid structures that move but maintain their structures when the water container is tilted. The density of the EPS-bound water resists minor hydraulic challenges. The definitions of all other consorms within cluster 2F are not based on the binding of free liquid water by EPS to define the consorm. A boil (2, 13-10 BOL) is a subcutaneous infection of an animal; it causes disruption of the tissues and releases of daughter water and possibly fermentation gases confined in tissues until relieved by the breakdown of surrounding tissues. A similar consorm is necrosis (2, 18-16 NCR). NCR causes rotting of tissues, causing cavitation and collapse of infested cells, leaving an infective process that continues to degrade surrounding tissues.

Rot consorms (2, 14-15 ROT) inflict more generalized corrosive attacks, more commonly seen in plants. The infested tissues rot quickly and cause localized and

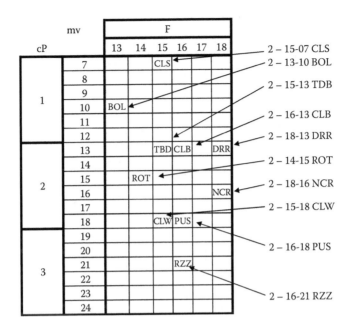

FIGURE 10.5　Grid positioning of alpha two consormial clustering 2F.

then generalized collapse of the host. Coliform and diarrhea consorms (2, 16-13 CLB and 2, 18-13 DRR, respectively) infest animal intestines. CLB dominated by coliform bacteria is generally nonsymptomatic. DRR consorms generally infest the deeper regions of the intestinal tract and disrupt the normal synthesis and evacuation of solid excreta. DRR is particularly active in the binding of water in EPS at the expense of the host's water requirements. Although the fecal material has a high water content, the host cannot maintain a healthy water balance and suffers from dehydration.

The remaining 2O consorm is unusual in that it forms complex interrelationships with plant roots in a common habitat. Rhizosphere (2, 16-21 RZZ) is complex; it contains a broad range of mycelia and other bacteria and fungi. The intelligence and intraconsormial organization of RZZ allow interactions with plant roots involving movements of water, nutrients, and oxygen across the root: soil barrier that benefit both the plant and the consorm. An individual RZZ growing around roots can dominate the soil voids right out to the edge of the root zone. For a giant oak tree, the RZZ would spread outward in an area as wide as the roots. RZZ is a vital but little understood consorm that combines complexity in function with simplicity in purpose (supporting soil fertility and plant growth).

Figure 10.6 shows clustering in 2GR. Its 13 consorms thrive on the reduction side of the ORP interface with 06 to 12 fmv; water viscosity ranges from 07 to 24 fcP. Most consorms relate to natural or infectious processes in animals (2, 12-07 URN = urine; 2, 10-08 KDS = kidney stones; 2, 10-12 FCB = fecal coliform bacteria; 2, 11-14 CCD = carietic condition; 2, 12-17 ABS = abscess; 2, 11-12 CHR = cholera; 2, 06-19 ULC = ulcer; 2, 08-20 CBC = carbuncle; and 2, 07-23 FEC = feces). One consorm (2, 06-24 ROE) relates to the spoilage of eggs, one to plant diseases (2, 12-13 WLT); and the final (2,09-15 MIC) relates to the initial microbiological events associated with corrosion.

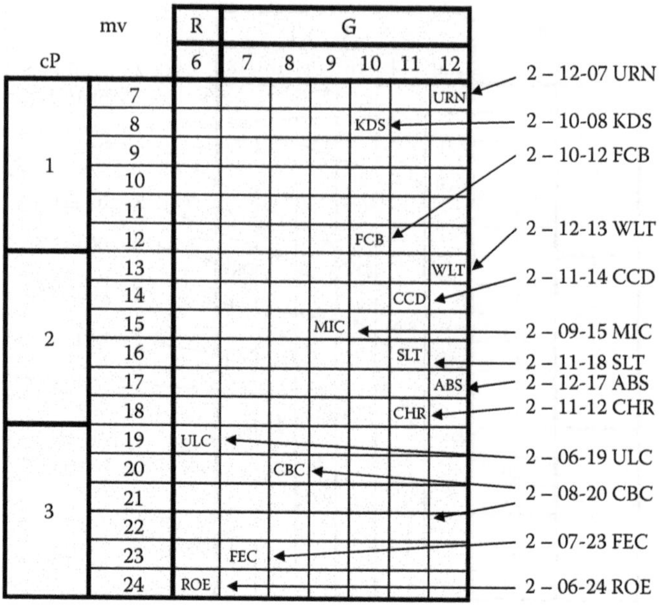

FIGURE 10.6   Grid positioning of alpha two consormial clustering 2GR.

Two consorms are associated with infections in the urinary tract (2, 12-07 URN) and the formation of concretious kidney stones (2, 10-08 KDS). Several other bacterial consorms are active in the intestine. One (2, 07-23 FEC) conserves body water by allowing only releases tightly bound water in the feces; freer liquid water is returned to the body. Dysfunction in this consorm leads to failure to recycle water and diarrhea (2, 18-13 DRR). Vibrioid bacteria dominate one consorm (2, 11-18 CHR) and cause rapid dehydration and cholera. Fecal coliform bacteria (2, 10-12 FCB) inhabit the intestine and serve as markers for hygiene risks since their presence in water or food indicates possible fecal contamination. Localized infectious processes can also cause intestinal dysfunction and are associated with abscesses (2, 12-17 ABS), ulcers (2, 06-19 ULC), and carbuncles (2, 08-20 CBC). Teeth and gums can also be infested with a carietic condition (2, 11-14 CCD) that leads to gum damage and loss of teeth.

Other 2GR consorms create other concerns. The avian egg is probably the most effectively packaged products in Nature. It remains effectively sterile while providing a secure environment for a growing embryo. The rotten egg consorm (2, 06-24 ROE) can breach an egg shell. After it penetrates the membranous defenses and resists the biocidal lysozymes in eggs, the bacteria cause spoilage. The wilt consorm (2, 12-13 WLT) in association with fungi, can impair the ability of plants to maintain adequate hydrostatic pressures in tissues, resulting in wilt. Most problems are generated by infestations in the xylem and phloem (conductive tissues) of plants.

The final cluster in 2GR is the broad spectrum of bacteria involved in microbiologically influenced corrosion (2, 09-15 MIC). MIC is a complex consorm that adapts to local environmental conditions. The first stage is growth of a biomass (often as a biofilm) on a surface vulnerable to electrolytic or adidulolytic corrosion. Complex bioconcretions within the biomass effect protection for the biomass and cause corrosion of underpinning surfaces (see also corrosion, 3, 06-16 CRS). Corrosion was considered a physico-chemical event arising from electrolytic destruction by hydrogen sulfide of metal alloys consisting mainly of iron. Corrosion is the process of corroding (e.g., rusting) and the damages resulting from corrosive processes. Lost production, secondary environmental damages, and expenses for replacement, regeneration, and preventative maintenance costs arising from corrosion total billions of dollars.

10.5 ALPHA THREE: INORGANIC BIOCONCRETING CONSORMS

Alpha three bacterial consorms generate organic carbon-based biomass, usually under reductive conditions or at the oxidation–reduction interface. That activity leads to the accumulation of inorganic elements (iron, manganese, aluminum, and under some conditions, carbonates) beyond metabolic needs of the consorm. Figure 10.7 is the grid atlas for alpha three consorms. Cluster 3R (Figure 10.8) includes very reductive consorms. Cluster 3GF (Figure 10.9) generally operates near or at the oxidation–reduction interface (fmv 7 to 22) and 03 to 30 fcp. Cluster 3O (Figure 10.10) shows the five consorms that operate oxidatively (fmv 19 to 24).

Nineteen consorms in alpha three are preliminarily differentiated into clusters 3R (five consorms), 3GF (nine consorms). and 3O (five consorms). Cluster 3R incorporates three consorms directly related to microbiological corrosion processes emanating from 2, 09-15 MIC. Generalized corrosion (3, 06-16 CRS) structurally damages alloys and concrete but has no concentrated site of activity. Focused corrosion usually

cP	mv	R 1	2	3	4	5	6	G 7	8	9	10	11	12	F 13	14	15	16	17	18	O 19	20	21	22	23	24
0	1																								
	2																								
	3																			CRB					
	4																								
	5																								
	6																								
1	7																								
	8																								
	9																								
	10									SBL															
	11																								
	12																								
2	13																								
	14																			LSL					
	15																								
	16			BWR		CRS																			
	17															TCL									
	18																						IPN		
3	19			PFR																OCR					
	20																								
	21										BPL			III						PLG					
	22																								
	23																								
	24					PTG																			
4	25																			CCR					
	26																			RST					
	27				CGG						BBR														
	28															MGN				FRD					
	29																								
	30										PGI														

FIGURE 10.7 Grid atlas differentiation for 3, 13-21 alpha three consormial clusters.

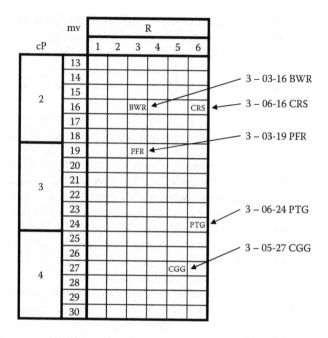

FIGURE 10.8 Grid positioning of alpha three consormial clustering 3R.

starts with pitting (3, 06-24 PTG) that terminates with structural perforations (holes) created by 3, 03-19 PFR). Pitting can cause structural failure and collapse; perforation is site-specific.

Two other reductive 3R cluster events produce very different phenomena. Black waters (3, 03-16 BWR) are generated in very reductive conditions by a surfeit of iron sulfides and carbonates. Another form of water plugging arises from the binding of biomass to sands, silts (see 2, 11-16 SLT), and clays into a matrix to create clogging and prevent groundwater movement (3, 05-27 CGG).

Cluster 3GF includes nine consorms associated with gradients around the oxidation–reduction interface. Two relate to lateral layers of biomass created in porous media often associated with the static water level. One consorm forms a black plug layer (3, 10-21 BPL) that becomes relatively impermeable to water; lateral slime layers (3, 18-14 LSL) are loosely associated within porous media. In general, the biomasses form non-specific concretious growths (3, 18-25 CCR). A biomass capped by a protective iron-rich concretion is a tubercle (3, 15-17 TCL). A poorly defined iron-rich concreted material in a biomass is ochrous (3, 18-19 OCR). On occasion, bioaccumulated iron or manganese may be layered in rusticles (3, 19-26 RST). The manganese in the layers may create ball-like structures known as manganese nodules (3, 15-18 MGN).

Plugging consorms can form in sands. The formations appear to "boil" when they relieve excessive water pressures created by the biomass plugs diverting the water upward. This phenomenon is caused by sand boils (3, 10-10 SBL). An associated activity of rusticles under more reductive conditions is the formation of iron-rich spheres 1 to 30 microns in diameter. Called "blueberries" (3, 10-17 BBR), the spheres constitute other sources of structural damage to steel pipelines by gouging the walls

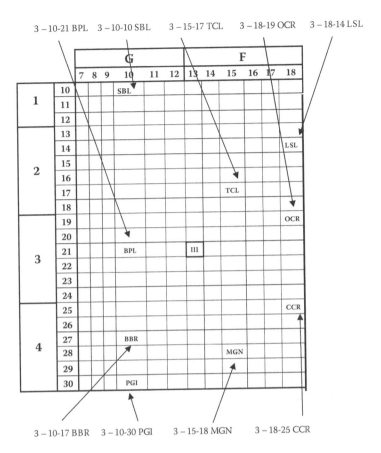

FIGURE 10.9　Grid positioning of alpha three consormial clustering 3GF.

and causing leakage. The gouging is more common in the floors of pipes because of the particle density.

When a rusticle matures under more reductive conditions accompanied by saturation with ferric iron, the biomass becomes less active and the spent rusticle takes on the chemistry of pig iron (3, 10-30 PGI). Figure 10.10 illustrates cluster 3O. It contains five bacterial consorms involved in the oxidative accumulation of inorganics. The one that synthesizes carbonates (3, 19-03 CRB) is different from the others concerned with ferric iron accumulations. Carbonates are synthesized in oxidative environments and are commonly formed in microbial fuel cells. Activities of the other four consorms focus on ferric iron as a product of biomass generated by rusticles (3, 19-16 RST) or plugging (3, 19-21 PLG), leading at maturation to the formation ferric iron-rich pans (3, 22-18 IPN) or ferric-rich deposits (3, 20-28 FRD). The four ferric-rich consorms have a geological connection.

10.6　ALPHA FOUR: CARBON REDUCING CONSORMS

One inevitable consequence of biosphere activities is the generation of a bulk mass of organic carbon mainly in the surface biosphere. This pool of organic carbon

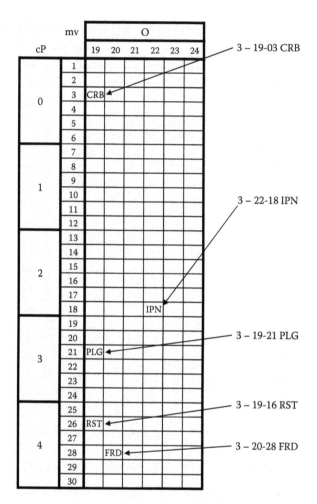

FIGURE 10.10   Grid positioning of alpha three consormial clustering 3O.

moves downward via gravity from the oxidative surface biosphere into the increasing reductive and nutrient-depleted crust of the Earth. When organic carbon forms enter water-saturated parts of the crust, bacteria in these reductive regions use the descending organic matters as sources of energy and nutrients. Alpha five consorms usually undertake these functions. Energy and nutrients are extracted, leaving molecules dominated by carbon and hydrogen as of crude oils, gases, or coals. Only one alpha four cluster participates in this reduction of organic carbon (Figure 10.11).

Organic carbon is essentially reduced to combustible forms of carbon and hydrogen. The initial process is stripping all nutrients including oxygen. One major terminal product of this respiratory function is the production of carbon dioxide. At the same time, nitrogen, phosphorus, sulfur, magnesium, potassium and the other nutritive elements are removed by the alpha four consorms. This leaves a number of recalcitrant pooled products. Carbon dioxide is reduced as a primary product of respiration. The removal of oxygen from carbon dioxide by consorm 4, 04-20

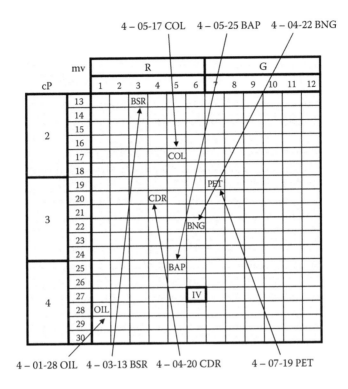

FIGURE 10.11   Grid atlas differentiation for 3, 13-21 alpha four consorms.

CDR provides feedstocks of reduced carbon. If the carbon undergoes reduction to elemental carbon, the conform involved would be 4, 05-17 COL. Coal is generated in a lateralized or crystallized form, perhaps by the ability of these bacteria to also remove hydrogen.

When biocrystallization to black coal is incomplete, consorm 4, 07-19 PET forms intermediate, highly combustible, carbon-rich products such as peat and lignite. When the organic carbon is reduced to complex polymeric (petroleum) hydrocarbons, consorm 4, 01-28 OIL is responsible for depositing oil reserves under very reductive conditions in water-saturated regions of the Earth's crust. The reduction of organic carbon terminates when a maximum number of hydrogen ions associates with a single carbon atom and forms methane (CH_4) produced by 4, 04-22 BNG.

When crude oil pooled deep in natural reservoirs is extracted, a bacterial consorm will "mine" the resident water from the oil to initiate metabolism through assimilation and electrolysis. To protect this newly mined water, the 4, 05-25 BAP consorm generates black asphaltene coatings. The biomass becomes dense with black "goop" and attaches to pipeline surfaces. This reduces the transmissivity of oil, increases pumping costs, and imposes risks of corrosion. BAP may contain high populations of aerobic (oxidative) bacteria based on the ability of the bacteria to use electrolytic charges to break down the accumulated water into oxygen and hydrogen. Another alpha four consorm generates black smokers (4, 03-13 BSR). These bacteria are found in great numbers at the extremely active gradients formed by waters rising as vents from the deep oceans. These waters carry many hydrocarbons that serve as major carbon sources and foster growth of the biomass and secondarily the abundant animal growth found around the vents.

10.7 ALPHA FIVE: CARBON OXIDIZING CONSORMS

Alpha five exhibits overlaps with alpha two consorms because their function is to oxidize reduced forms of carbon as hydrocarbons when the hydrocarbons re-emerge in the oxidative zones of the surface biosphere (Figure 10.12). Transient movements of these hydrocarbons upward into oxidative zones utilizes covert diffusion (volatile hydrocarbons), seepage of lighter density hydrocarbonaceous fluids (crude oils), and the venting of hydrocarbon gases (methane). These forms of hydrocarbon serve as energy sources for bacteriological activities after they enter the reduction–oxidation interfaces beneath the biosphere. Figure 10.13 shows the various bacterial consorms that oxidatively assimilate reduced hydrocarbons.

Only the gas hydrate consorm (3, 15-10 GHY) works alone as part of alpha five. Gas hydrates are also known as clathrates and occur at the sediment:water interfaces in deep-ocean environments, particularly on continental shelves. They appear ice-like, contain disproportionately large amounts of methane (water:methane, 1:8), and function above the freezing point of water. The hydrates are not well understood because they are unstable although they represent large potentially exploitable energy reserves. Gas hydrates have been found under frozen land masses (e.g., tundra). The only bacteriology known is that the capping of the hydrate appears to be dominated by bacteria that generate a type of EPS that acts as a surfactant. The surfactant forms an efficient frozen water (ice) web within which methane molecules cluster in a stable manner under reductive conditions. It is possible that 1, 16-12 ICE is a major consorm participating in gas hydrate because ice bacteria can "grow" ice at temperatures as high as 7°C.

Most bacterial consorms involved in oxidative degradation of hydrocarbons around the oxidation–reduction interface: seven at the front (F), four at sites on the reductive side (G or R), and four on the oxidative (O) side. The integrated activities of the alpha five, alpha one, and alpha two consorms complete carbon cycle in the surface and subsurface biospheres. Essentially, organic carbon reduces to hydrocarbons and elemental carbon and the other elements are stripped off. As the stripped forms of carbon reemerge into the surface biosphere, they are oxidatively degraded via the synthesis of organics associated with life and the production of carbon dioxide.

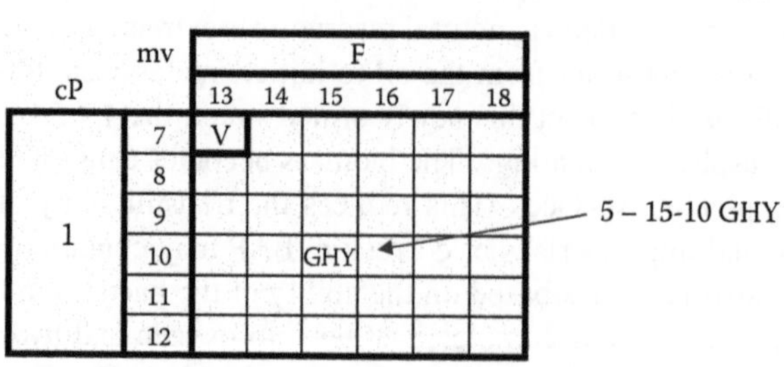

FIGURE 10.12 Grid atlas differentiation for 5, 13-07 alpha five consorms.

cP	mv	1	2	3	4	5	6	7	8	9	10	11	12	13	14	15	16	17	18	19	20	21	22	23	24
		R						G						F						O					
1	7															CLS			BBL				SHN		
	8																								
	9																								
	10										SBL														
	11			BSR													ICE						LMN		
	12															TBD			LSL						TCT
2	13																								
	14											SLT													
	15									MIC															
	16															TCL									
	17															CLW							IPN		
	18																								

FIGURE 10.13 Bacterial consorms primarily from alpha one and two groups considered capable of degrading hydrocarbons entering oxidative surface biosphere.

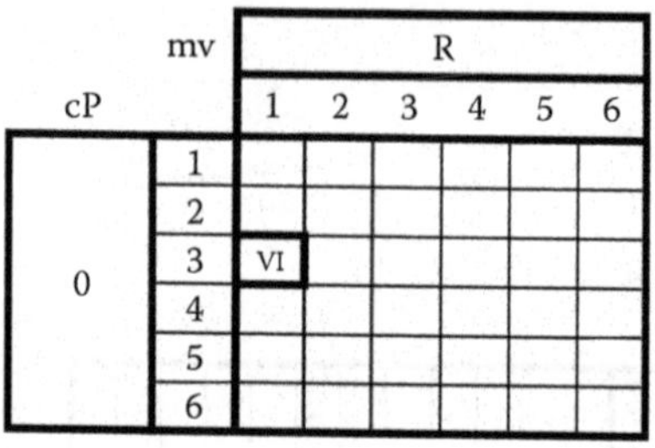

cP	mv	R					
		1	2	3	4	5	6
0	1						
	2						
	3	VI					
	4						
	5						
	6						

FIGURE 10.14 Grid atlas differentiation for 6, 01-03 alpha six grouping.

A carbon cycle generated between the oxidative surface biosphere and the reductive subsurface biosphere causes the carbon to recycle. In this sense, oil and gas are renewable sources of energy (with very long recycling times) fed by the entry of organic materials into the reductive zones. Through the natural recycling of carbon enters the reduced (oil and gas) state. Via seepage, leakage, and diffusion) it reenters the oxidative surface biosphere and thus recycles.

10.8 ALPHA SIX: HYPERBARIC DISPERSED BIONUCLEATING CONSORMS

Science has always faced the unknown; for zoologists, the challenge has been understanding extreme forms of animals such as dinosaurs. Botanists examine the immense grandeur of the rainforests and the enormous diversity of oceanic phytoplankton. The challenge for bacteriologists may be exploring the extreme outer limits of life in the subsurface biosphere and in biospheres on other celestial bodies.

The alpha six group (6. 01-13 HBC) remains unexplored. It occupies an extreme environment—high pressures and temperatures and water cycling between liquid and gaseous states (see Figure 10.14). This environment has been active since water first cooled the crust enough to penetrate it. Alpha six consorms may fulfill the dream of a bacteria hunter to find perhaps the earliest forms of active life on Earth. The search would require descent into the depths of Earth's the crust or replicating the extreme conditions there in a laboratory in the surface biosphere.

10.9 DIFFERENTIATION OF MAJOR CONSORMS BY GRID POSITIONS AND BART REACTIONS

Characterization of the consorms is based on the three-letter code assigned, the factorial ranges of ORP and water viscosity, the type of cluster to which the consorm belongs, the natural pH and temperature ranges required activity, the normal color range observed, and finally the potential reaction type and time lapses generated by the HAB, SRB, IRB, SLYM, DN, and APB BART analyses. A focal point location (FPL) is applied to each consorm and the six major consormial groups. Each consorm functions effectively over a range of conditions defined on the grid. Tables 10.1 through 10.16 define the characteristics in groups of four consorms; thus the tables characterize 64 consorms in order of cluster types.

TABLE 10.1
Characterization of Group 1 Consorms (CLD, CMT, FOM, and ICE)

FPL	1, 22-03	1, 17-15	1, 19-06	1, 16-12
Code	CLD	CMT	FOM	ICE
fmv range	16–24	5–24	15–22	15–22
fcP range	1–6	7–24	6–12	11–24
Cluster type	1	1	1	1
pH range	3.5–9.5	NK	2.5–10.5	5.5–8.5
Temperature range	0–50°C	−25 to 2°C NK	0 to 40°C	−15 to 8°C
Color	Clear to grey	White to grey	Clear to beige	White
HAB reaction	UP	NK	UP	UP
HAB TL range	0.5–4 days	NK	1–5 days	2–6 days
SRB reaction	BT	NK	ND	ND
SRB TL range	5 – ND	NK	ND	ND
IRB reaction	ND	NK	CL, FO, or ND	ND
IRB TL range	ND	NK	3 – ND	ND
SLYM reaction	CL (TH)	NK	CL	CL
SLYM TL range	0.5–3 days	NK	0.5–4 days	1–4 days
DN reaction	ND	NK	FO or ND	ND
DN TL range	ND	NK	3–5 or ND	ND
APB reaction	ND	NK	ND	ND
APB TL range	ND	NK	ND	ND

TABLE 10.2
Characterization of Groups 1 and 2O Consorms (LNG, SNW, BCH, and SHN)

FPL	1, 16-03	1, 19-12	2, 19-07	2, 22-07
Code	LNG	SNW	BCH	SHN
fmv range	NK	17–24	7–18	19–24
fcP range	NK	7–18	6–21	7–12
Cluster type	1	1	20	20
pH range	NK	5.5–8.5	5.5–9.5	6.5–10.5
Temperature range	>100°C	−15 to 2°C	30–46°C	5–35°C
Color	Bright white	White	White to grey	Iridescent
HAB reaction	NK	UP	UP	UP
HAB TL range	NK	3–8 days ND	0.5–3 days	0.5–4 days
SRB reaction	NK	ND	ND or BT	ND or BT
SRB TL range	NK	ND	ND or >5 days	ND or >5 days
IRB reaction	NK	ND	ND or CL	ND or CL
IRB TL range	NK	ND	ND or >2 days	ND or >3 days
SLYM reaction	NK	CL	CL (TH)	CL (TH)
SLYM TL range	NK	2–5 days ND	0.5–2 days	0.5–2 days
DN reaction	NK	ND	ND	ND
DN TL range	NK	ND	ND	ND
APB reaction	NK	ND	ND (DY)	ND
APB TL range	NK	ND	ND (>4 days)	ND

TABLE 10.3

Characterization of Group 2O Consorms (LMN, TCT, PEY, and PLG)

FPL	2, 22-12	2, 24-13	2, 23-14	2, 19-21
Code	LMN	TCT	PEY	PLG
fmv range	16–24	19–24	16–24	4–21
fcP range	7–18	7–18	10–18	13–24
Cluster type	20	20	20	20
pH range	6.5–9.5	07–Sep	07–Sep	4–12.5
Temperature range	0–35°C	15–35°C	30–42°C	–05–35°C
Color	Light blue	None	Pink–red	Grey to brown
HAB reaction	UP	UP	UP	UP
HAB TL range	0.5–3 days	0.5–2 days	0.5–2 days	1–3 days
SRB reaction	ND	ND	ND	BT
SRB TL range	ND	ND	ND	>2 days
IRB reaction	ND	ND or CL	ND	CL (FO)
IRB TL range	ND	ND or >4 days	ND	>3 days
SLYM reaction	CL	CL	CL (TH)	CL (SR)
SLYM TL range	0.5–2 days	0.5–2 days	0.5–2 days	0.5–2 days
DN reaction	ND	ND	ND	FO (ND)
DN TL range	ND	ND	ND	>3 days (ND)
APB reaction	ND	ND	ND	ND (DY)
APB TL range	ND	ND	ND	ND (>3 days)

TABLE 10.4

Characterization of Groups 2O and 2F Consorms (PMP, SKN, SLM, and CLS)

FPL	2, 22-21	2, 23-21	2, 21-22	2, 15-07
Code	PMP	SKN	SLM	CLS
fmv range	13–24	19–24	18–24	13–21
fcP range	15–24	17–24	13–24	6–14
Cluster type	20	20	20	2F
pH range	6.5–9	5.5–10.5	3.5–12.5	6.5–10.5
Temperature range	34–44°C	5–35°C	–05–35°C	5–35°C
Color	Pink–yellow	Grey to brown	White–grey	Diffuse turbid white
HAB reaction	UP	UP	UP	UP
HAB TL range	0.5–2 days	0.5–4 days	0.5–4 days	0.5–5 days
SRB reaction	ND (BT)	ND	ND (BT)	ND (BT)
SRB TL range	ND or >4 days	ND	ND or >2 days	ND or >3 days
IRB reaction	ND	ND (CL)	CL (FO)	ND (CL or FO)
IRB TL range	ND	ND (>3 days)	>3 days	>3 days
SLYM reaction	CL	CL (SR)	CL (SR or TH)	CL (SR or TH)
SLYM TL range	0.5–2 days	0.5–3 days	0.5–2 days	1–3 days
DN reaction	ND	ND	ND (FO)	ND
DN TL range	ND	ND	ND (>3 days)	ND
APB reaction	ND (DY)	ND (DY)	ND (DY)	ND
APB TL range	ND (>4 days)	ND (>4 days)	ND (>3 days)	ND

TABLE 10.5
Characterization of Group 2F Consorms (BOL, TBD, CLB, and DRR)

FPL	2, 13-10	2, 15-13	2., 16-13	2, 18-13
Code	BOL	TBD	CLB	DRR
fmv range	7–18	3–24	4–18	3–21
fcP range	7–15	9–15	9–21	10–21
Cluster type	2F	2F	2F	2F
pH range	5.5–9.5	4–12.5	6.5–9.5	5.5–8.5
Temperature range	34–44°C	–05–45°C	24–45°C	34–44°C
Color	Yellow gel	Dense turbid white	Grey	Watery grey
HAB reaction	UP	UP	UP	DO or UP
HAB TL range	0.5–3 days	0.5–3 days	0.5–3 days	0.5–2 days
SRB reaction	ND	ND	ND (BT)	BT (BB)
SRB TL range	ND	ND	ND or >3 days	> 1 day (>2 days)
IRB reaction	ND	ND	ND (CL, or FO)	FO CL
IRB TL range	ND	ND	>1 day	>2 days
SLYM reaction	CL (TH)	CL (SR or TH)	CL (FO or TH)	CL (FO)
SLYM TL range	0.5–2 days	0.5–3 days	0.5–2 days	0.5–2 days
DN reaction	ND	ND (FO)	ND (FO)	ND
DN TL range	ND	ND (>3 days)	ND (>1 day)	ND
APB reaction	ND (DY)	ND (DY)	ND (DY)	DY
APB TL range	ND (>4 days)	ND (>2 days)	ND (>1 day)	>2 days

TABLE 10. 6
Characterization of Group 2F Consorms (ROT, NCR, CLW, and PUS)

FPL	2, 14-15	2, 18-16	2, 15-18	2, 16-18
Code	ROT	NCR	CLW	PUS
fmv range	10–18	4–21	13–18	7–21
fcP range	10–22	13–18	16–24	11–24
Cluster type	2F	2F	2F	2F
pH range	5.5–8.5	6.5–9.5	4–12.5	6.5–9.5
Temperature range	5–35°C	5–44°C	2–45°C	34–44°C
Color	Viscid grey	Darkening crust	Dense white gel	Viscid yellow
HAB reaction	DO or UP	UP	UP	UP
HAB TL range	0.5–4 days	0.5–4 days	0.5–3 days	0.5–2 days
SRB reaction	BT (BB)	ND (BT)	BT	ND (BT)
SRB TL range	>2 days (>3 days)	ND or >1 day	>3 days	ND or (>1 day)
IRB reaction	CL	ND (CL, or FO)	ND	ND
IRB TL range	>4 days	(>3 days)	ND	ND
SLYM reaction	CL (FO)	CL (FO or TH)	CL	CL (FO or TH)
SLYM TL range	0.5–4 days	0.5–3 days	0.5–3 days	0.5–3 days
DN reaction	ND	ND (FO)	ND (FO)	ND
DN TL range	ND	ND (>3 days)	ND (>3 days)	ND
APB reaction	DY	ND (DY)	ND (DY)	DY
APB TL range	>2 days	ND (>2 days)	ND (>2 days)	>2 days

TABLE 10.7

Characterization of Groups 2F and 2GR Consorms (RZZ, URN, KDS, and FCB)

FPL	2, 16-21	2, 12-07	2, 10-08	2, 10-12
Code	RZZ	URN	KDS	FCB
fmv range	3–22	9–15	7–12	4–18
fcP range	13–31	6–15	7–14	9–21
Cluster type	2F	2GR	2GR	2GR
pH range	4.5–10.5	4.5–9.5	5.5–10.5	6.5–9.5
Temperature range	−05–55°C	34–46°C	34–46°C	24–45°C
Color	Grey	Clear to yellow	White	White
HAB reaction	DO or UP	UP	UP	UP
HAB TL range	0.5–4 days	0.5–6 days	0.5–6 days	0.5–2 days
SRB reaction	BT (BB)	ND (BT)	ND	ND (BT)
SRB TL range	>3 days (>4 days)	ND (>4 days)	ND	ND or >4 days
IRB reaction	FO CL	ND	ND	ND
IRB TL range	>2 days	ND	ND	ND
SLYM reaction	CL (FO)	CL (FO)	CL	CL (FO or TH)
SLYM TL range	0.5–3 days	0.5–4 days	0.5–6 days	0.5–2 days
DN reaction	FO	ND	ND	ND
DN TL range	>3 days (>4 days)	ND	ND	ND
APB reaction	DY	ND (DY)	ND	DY
APB TL range	>2 days	ND (>2 days)	ND	>2 days

TABLE 10.8

Characterization of Group 2GR Consorms (WLT, CCD, MIC, and SLT)

FPL	2, 12-13	2, 11-14	2, 09-15	2, 11-16
Code	WLT	CCD	MIC	SLT
fmv range	6–18	3–16	3–19	7–20
fcP range	9–18	13–21	9–31	13–30
Cluster type	2GR	2GR	2GR	2GR
pH range	5.5–9.5	3.5–8.5	1.5–10.5	3.5–10.5
Temperature range	2–30°C	34–44°C	−05–55°C	−05–55°C
Color	Viscid grey	Viscid grey	Brown–black	Grey
HAB reaction	UP	UP	UP	UP (DO)
HAB TL range	1–5 days	0.5–3 days	0.5–4 days	0.5–5 days
SRB reaction	BT (BB)	BT (BB)	BT (BB)	BT (BB)
SRB TL range	>3 days (>4 days)	>3 days (>4 days)	>3 days (>4 days)	>3 days (>4 days)
IRB reaction	ND	ND	FO (CL)	FO (CL)
IRB TL range	ND	ND	>2 days	>2 days
SLYM reaction	CL	CL	CL	CL
SLYM TL range	1–4 days	0.5–4 days	0.5–4 days	0.5–5 days
DN reaction	ND	ND	ND	ND (FO)
DN TL range	ND	ND	ND	ND >3 days
APB reaction	ND	DY	DY	DY
APB TL range	ND	>2 days	>1 day	>2 days

TABLE 10.9
Characterization of Group 2GR Consorms (ABS, CHR, ULC, and CBC)

FPL	2, 12-17	2, 11-12	2, 06-19	2, 08-20
Code	ABS	CHR	ULC	CBC
fmv range	10–21	6–20	4–12	2–12
fcP range	13–24	6–16	13–24	16–30
Cluster type	2GR	2GR	2GR	2GR
pH range	5.5–9.5	6.5–9.5	3.5–8.5	2.5–8.5
Temperature range	34–44°C	34–44°C	32–46°C	28–44°C
Color	Viscid grey	Grey	Grey	Concreted grey
HAB reaction	UP (DO)	UP (DO)	UP	DO (UP)
HAB TL range	0.5–5 days	0.5–2 days	0.5–3 days	0.5–3days
SRB reaction	BT	BT	ND (BT)	ND (BT)
SRB TL range	>4 days	>2 days	ND (>3 days)	ND (>4 days)
IRB reaction	ND	ND	ND	ND
IRB TL range	ND	ND	ND	ND
SLYM reaction	CL (TH or DS)	CL (TH or DS)	CL (DS)	CL (DS)
SLYM TL range	0.5–4 days	0.5–2 days	0.5–2 days	0.5–2 days
DN reaction	ND (FO)	ND	ND	ND
DN TL range	ND >3 days	ND	ND	ND
APB reaction	DY	DY	DY	DY
APB TL range	>3 days	>1 day	>1 day	>0.5 day

TABLE 10.10
Characterization of Groups 2GR and 3R Consorms (FEC, ROE, BWR, and CRS)

FPL	2, 07-23	2, 06-24	3, 03-16	3, 06-16
Code	FEC	ROE	BWR	CRS
fmv range	3–12	3–12	13–20	1–15
fcP range	19–27	13–26	1–12	13–24
Cluster type	2GR	2GR	3R	3R
pH range	3.5–9.5	4.5–8.5	3.5–9.5	1.5–11.5
Temperature range	32–46°C	−5–44°C	−2–40°C	−15–50°C
Color	Grey brown	Grey–red–black	Grey–black	Grey–black
HAB reaction	DO (UP)	UP (DO)	DO (UP)	DO (UP)
HAB TL range	0.5–2 days	0.5–8 days	0.5–2 days	0.5–2 days
SRB reaction	ND (BT)	ND (BT)	BT (BB)	BB (BT)
SRB TL range	ND (>1 day)	ND (>1 day)	>1 day	0.5–5 days
IRB reaction	ND	ND	FO	FO CL DG BC
IRB TL range	ND	ND	1–3 days	0.5–4 days
SLYM reaction	CL (DS)	CL (DS)	CL (DS)–BL	CL (DS)–BL
SLYM TL range	0.5–2 days	0.5–4 days	0.5–4 days	0.5–4 days
DN reaction	ND	ND	ND	ND
DN TL range	ND	ND	ND	ND
APB reaction	DY	DY	DY	DY
APB TL range	>1	>2	1–8 days	1–8 days
Iron percent dry wt			0.1–7%	7–75%

TABLE 10.11

Characterization of Groups 3R and 3GF Consorms (PFR, PTG, CGG, and BPL)

FPL	3, 03-19	3, 06-24	3, 05-7	3, 10-21
Code	PFR	PTG	CGG	BPL
fmv range	1–9	3–15	1–18	4–18
fcP range	13–30	19–30	19–30	15–27
Cluster type	3R	3R	3R	3GF
pH range	3–9.5	4.5–11	5.5–12.5	3–9.5
Temperature range	–2–40°C	–5–50°C	–5–35°C	12–44°C
Color	Brown–black	Brown	Beige–brown	Black
HAB reaction	DO (UP)	UP	UP (DO)	UP (DO)
HAB TL range	0.5–2 days	0.5–4 days	1–6d	1–4 days
SRB reaction	BB (BT)	BT (BB)	BT (BB)	BT (BB)
SRB TL range	0.5–8 days	1–8 days	1–14 days	1–14 days
IRB reaction	FO CL	FO CL DG	CL DG FO BR	CL BC BR BL
IRB TL range	2–7 days	1–6 days	2–8 days	1–8 days
SLYM reaction	CL (DS)–BL	CL (DS)–BL	CL (DS)–BL	CL (DS)–BL
SLYM TL range	0.5–4 days	0.5–4 days	0.5–6 days	0.5–6 days
DN reaction	ND	ND	ND	ND (FO)
DN TL range	ND	ND	ND	ND (2–6 days)
APB reaction	DY	DY	DY	DY
APB TL range	1–8 days	1–8 days	1–8 days	1–8 days
Iron percent dry wt	20–80%	30–85%	0.5–40%	0.5–10%

TABLE 10.12

Characterization of Group 3GF Consorms (TCL, OCR, LSL, and BBR)

FPL	3, 15-17	3 , 18-19	3, 18-14	3, 10-27
Code	TCL	OCR	LSL	BBR
fmv range	9–15	13–24	10–22	7–21
fcP range	7–22	10–24	7–18	19–30
Cluster type	3GF	3GF	3GF	3GF
pH range	3–10.5	6–12.5	4.5–11.5	5.5–9.5
Temperature range	2–60°C	2–40°C	6–45°C	0–30°C
Color	Brown	Brown	Layered greys	Black
HAB reaction	UP (DO)	UP (DO)	UP (DO)	ND (UP or DO)
HAB TL range	1–6 days	1–8 days	1–4 days	ND (1–8 days)
SRB reaction	BT (BB)	ND (BT or BB)	BT or BB	ND (BT or BB)
SRB TL range	2–10 days	ND (2–14 days)	2–14 days	ND (2–14 days)
IRB reaction	CL BC BR BL	CL BG BC	CLFO	ND (CL FO)
IRB TL range	1–8 days	1–8 days	2–10 days	ND (2–10 days)
SLYM reaction	CL (DS)–BL	CL (DS)	CL (DS BL)	ND (CL or DS)
SLYM TL range	0.5–6 days	0.5–6 days	0.5–8 days	ND (0.5–8 days)
DN reaction	ND	ND (FO)	ND (FO)	ND (FO)
DN TL range	ND	ND (4–7 days)	ND (4–7 days)	ND (4–7 days)
APB reaction	DY	DY	DY	ND (DY)
APB TL range	1–8 days	2–8 days	1–5 days	ND(1–5 days)
Iron percent dry wt	15–40%	20–85%	0.1–20.0%	0.1–98%

TABLE 10.13

Characterization of Groups 3GF and 3O Consorms (PGI, MGN, CCR, and CRB)

FPL	3, 10-30	3, 15-28	3, 18-25	3, 19-03
Code	PGI	MGN	CCR	CRB
fmv range	7–13	10–24	10–20	16–24
fcP range	24–30	22–30	19–27	1–18
Cluster type	3GF	3GF	3GF	3O
pH range	5.5–10.5	5.5–10.5	4.5–11.5	7–12.5
Temperature range	−10–40°C	−10–40°C	+2–40°C	+2–40°C
Color	Brown	Grey–black	Light grey–beige	White–grey
HAB reaction	ND (UP)	UP	UP	ND
HAB TL range	ND(1–8 days)	2–8 days	3–10 days	ND
SRB reaction	ND (BT)	ND (BT)	ND	ND
SRB TL range	ND (6–14 days)	ND (6–14 days)	ND	ND
IRB reaction	ND (CL FO)	ND (CL FO)	ND (CL FO)	ND
IRB TL range	ND (6–10 days)	ND (6–10 days)	ND (5–12 days)	ND
SLYM reaction	ND (CL or DS)	CL or DS	CL or DS	CL or DS
SLYM TL range	ND (2–8 days)	(1–8 days)	2–8 days	3–10 days
DN reaction	ND	ND	ND	ND
DN TL range	ND	ND	ND	ND
APB reaction	ND (DY)	ND (DY)	ND	ND
APB TL range	ND (4–8 days)	ND (4–8 days)	ND	ND
Iron percent dry wt	80–98%	1–20%	0.01–15%	0.01–1%

TABLE 10.14

Characterization of Group 3O consorms (IPN, PLG, RST, and FRD)

FPL	3, 22-18	3, 19-21	3, 19-26	3, 20-28
Code	IPN	PLG	RST	FRD
fmv range	16–24	10–24	7–21	13–22
fcP range	13–24	13–24	19–30	19–30
Cluster type	3O	3O	3O	3O
pH range	5.5–10.5	1.5–12.5	5.5–10.5	6.5–11.5
Temperature range	−2–32°C	−10–50°C	2–38°C	2–38°C
Color	Brown	Grey–brown	Brown	Red–brown
HAB reaction	UP	UP	UP (DO)	UP
HAB TL range	1–8 days	1–8 days	1–6 days (4–8 days)	1–8 days
SRB reaction	ND	ND (BT or BB)	ND (BT)	ND (BT)
SRB TL range	ND	ND (2–8 days)	ND (3–10 days)	ND (3–10 days)
IRB reaction	CL FO BG	CL FO BG	CL FO BG	CL FO BG
IRB TL range	3–12 days	3–12 days	3–12 days	3–12 days
SLYM reaction	CL or DS	CL–DS–(BL)	CL–DS–(SR)	CL–DS–(SR)
SLYM TL range	1–8 days	0.5–8 days	0.5–8 days	0.5–8 days
DN reaction	ND	ND	ND	ND
DN TL range	ND	ND	ND	ND
APB reaction	ND	ND (DY)	ND (DY)	ND
APB TL range	ND	ND(1–6 days)	ND (1–6 days)	ND
Iron percent dry wt	40–85%	0.01–20%	20–85%	30–85%

TABLE 10.15

Characterization of Group 4 Consorms (OIL, BSR, COL, and CDR)

FPL	4, 01-28	4, 03-13	4, 05-17	4, 04-20
Code	OIL	BSR	COL	CDR
fmv range	1–12	1–24	1–12	3–15
fcP range	25–30	7–18	13–24	13–26
Cluster type	4	4	4	4
pH range	2.5–9.5	1.5–11.5	5.5–11.5	4.5–11.5
Temperature range	–10–55°C	0–160°C	0–55°C	0–38°C
Color	Black	Black	Black	Grey–brown
HAB reaction	UP (DO)	UP (DO)	UP (DO)	UP (DO)
HAB TL range	1–10 days	0.5–4 days	2–10 days	2–12 days
SRB reaction	ND (BB)	BT	BT	BT
SRB TL range	ND (5–15 days)	1–8 days	4–14 days	4–14 days
IRB reaction	FO CL BG BL	FO CL BG	FO	CL BG
IRB TL range	3–15 days	2–15 days	4–15 days	4–15 days
SLYM reaction	CL–DS–BL	CL	CL	CL
SLYM TL range	0.5–10 days	0.5–10 days	0.5–10 days	0.5–10 days
DN reaction	ND	ND	ND	ND
DN TL range	ND	ND	ND	ND
APB reaction	ND (DY)	ND (DY)	ND (DY)	ND (DY)
APB TL range	ND (4–10 days)	ND (2–8 days)	ND (6–14 days)	ND (6–14 days)
Iron percent dry wt	0–0.5%	0–5%	0–5%	0–5%

TABLE 10.16

Characterization of Groups 4 and 5 Consorms (BAP, PET, BNG, and GHY)

FPL	4, 05-25	4, 07-19	4, 04-22	5, 15-10
Code	BAP	PET	BNG	GHY
fmv range	3–12	4–12	1–12	1–18
fcP range	21–28	15–22	16–28	7–15
Cluster type	4	4	4	5
pH range	4.5–11.5	2.5–6.5	3.5–9.5	3.5–9.5
Temperature range	0–45°C	–05–45°C	0–75°C	0–11°C
Color	Black	Brown–black	Grey	White–grey
HAB reaction	UP (DO)	UP (DO)	UP (DO)	UP (DO)
HAB TL range	2–12 days	1–8 days	1–8 days	0.5–6 days
SRB reaction	BT	BT (BB)	ND	ND
SRB TL range	4–14 days	6–14 days	ND	ND
IRB reaction	CL BG	ND	ND	ND
IRB TL range	4–15 days	ND	ND	ND
SLYM reaction	CL	CL	CL	CL DS
SLYM TL range	0.5–10 days	0.5–10 days	0.5–8 days	0.5–4 days
DN reaction	ND	ND	ND	ND
DN TL range	ND	ND	ND	ND
APB reaction	ND (DY)	DY	ND (DY)	ND (DY)
APB TL range	ND (6–14 days)	3–14 days	(2–9 days)	(2–9 days)
Iron percent dry wt	0–1%	0–1%	0%	0%

Cluster types are defined primarily by the number designation of the major consormial alpha group and secondarily by the location along the oxidation–reduction gradient where specific consorms may be active. The gradient location is expressed as the factorial millivolt range (fmv) and defines the nature of the habitat where activity occurs. Consorm growth is influenced by water viscosity, measured in log centipoise (fcP), as the second major characteristic determining position in the atlas grid. Water may be dispersed into aerosol droplets or at the other extreme bound tightly within an organic or inorganic matrix. The fmv is graded from 1 (most reductive) to 24 (most oxidative); the fcP is graded from 1 (most dispersed) to 30 (most bound or viscid). Table 10.17 relates the ORP and viscosity factorial values.

TABLE 10.17

Relationship of Factorial Millivolts (fmv), Oxidation–Reduction Potential (ORP, millivolts), Factorial Centipoise (fcP), and Log cP[a]

fmv	ORP (mv)	fcP	log cP
1	−600	1	0.2
2	−500	2	0.3
3	−440	3	0.4
4	−380	4	0.6
5	−320	5	0.7
6	−260	6	0.9
7	−200	7	1.0
8	−170	8	1.2
9	−140	9	1.4
10	−110	10	1.6
11	−80	11	1.8
12	−50	12	2.0
13	−35	13	2.0
14	−24	14	2.2
15	−13	15	2.4
16	−2	16	2.6
17	9	17	2.8
18	20	18	3.0
19	50	19	3.0
20	80	20	3.2
21	110	21	3.4
22	140	22	3.6
23	170	23	3.8
24	200	24	4.0
		25	4.0
		26	4.2
		27	4.4
		28	4.6
		29	4.8
		30	5.0

[a] *Centipoise values indicate viscosity of water.*

Cluster types are defined early in this chapter and are numbered from one to six based on alpha group characteristics. Cluster numbers start with the alpha group number. The next component is a one- or two-letter definition of the ORP group: R = reductive; G = reductive gradient; F = oxidation–reduction interface; and O = oxidative. Cluster types defined in this section are: 1 (no designation); 2 (designated 2O or 2GR), 3 (designated 3O, 3GF, or 3R), 4 (no designation); 5 (no designation); and 6 (no designation). Figure 10.13 defines the bacterial consorms from other groups that may participate in alpha five activities.

Two major physical environmental parameters affecting consormial activity are pH and temperature. Bacteria prefer specific ranges of pH and temperatures but individual consorms can often adapt and extend their ranges of activity. Determination of the pH range is very difficult to define with precision because most bacterial consorms are protected by bound water EPS or inorganic crystalline matrices. These matrix barriers protect incumbent cells from the natural environmental pH within which the consorm is active. This protection extends the tolerable pH range. Bacterial consorms can also buffer pH locally to maintain activity in pH-hostile environments. The pH range takes into account the potential for H buffering when the local environment becomes more acidic. When pH values are alkaline and exceed 9.5, the risk of concretion increases, particularly in an environment with sufficient calcium. The range takes into account the potential for a consorm to produce bioconcretious structures and remain active.

Temperature ranges represent maximum criteria under which a specific consorm can remain active. Some consorms such as 1, 16-12 ICE retain water in the liquid form even when the temperatures are below freezing. This "antifreeze" property is associated with the EPS (surfactants) produced by many bacteria. They cause the water to remain liquid and remain suitable as a solvent for bacterial activity. Gas hydrate (5, 15-10 GHY) consorms maintain activity in an inverse manner. Ice is formed at temperatures above 0°C (4 to 7°C) and acts as a methane accumulator, generating an 8:1 cluster ratio of methane to water molecules. Generally, consorms exhibit great ability to adapt to lower temperatures than optimum (ideal) temperatures and reduced ability to adapt to higher temperatures. The temperature range reflects this adaptation ability, assuming that the optimum is near the higher end of the range.

Colors remains challenging because they develop and change during the maturation of a consormial biomass. The descriptor defines the color prevalent at the time of maturation (maximum activity or maximum biomass). Some consormial descriptors cite two colors. The first indicates maximum activity and the second the largest biomass. In some consorms, the largest biomass may exhibit brown color as a result of the bioaccumulation of ferric iron. The tables in this chapter include the ranges of ferric iron accumulated within biomasses for alpha three, alpha four, and alpha five.

To establish the types of bacteria active in a consorm, six BART analyses define community activities (reactions) and ranges of activity time lapses (TLs). Patterns can indicate a single reaction (e.g., acid in APB BART and foam in DN BART tester), two reactions (e.g., DN or UP in HAB BART and BT or BB in SRB BART) to more complex reactions that generate replicable signatures for a particular consorm. SLYM BART is probably the most sensitive tester and clouding (CL) is the most common reaction. On occasion the tester indicates a significant primary or secondary reaction, usually slime thread formation (TH), a dense gel-like growth (DG) near the bottom of the tester, and formation of a slime ring (SR) around the BART ball. TH (thread-like) and DG reactions support the premise that slime forming bacteria generate tightly bound water within a defined biomass. Bacteria in a TH commonly string one to three distinctive threads between the BART ball at the top and the base of the tester. Time lapse photography of primary influents and effluents of aerated wastewater treatment plants reveal that these TH threads are very active even in HAB BART tester. While the TH reaction tends to involve vertical biomass, structures defined by the DG reaction occur when a dense gel-like growth appears in a tester. Slime growths are denser than the liquid medium column and form in the base (lower third) of the tester. Generally both TH and DG reactions terminate with a CL.

Oxidative (aerobic) and reductive (anaerobic) consorms are easily differentiated by both the HAB and SRB BART procedures. Aerobic activities generally support the UP and BT reactions while anaerobic dominance generates the DO and BB reactions.

IRB BART testers generate the most complex reaction sets based primarily on the oxidative (aerobic) or reductive (anaerobic) state of the environment tested. The tester reacts secondarily with the iron it contains to create different forms of bioaccumulated iron. Primary differentiation is based on the generation of clouding (CL, usually yellow) or foaming (FO, commonly with red-brown tinges) under aerobic or anaerobic conditions, respectively. Ferric-dominated secondary reactions reflect the aerobic nature of IRB in a consorm. They may start with a brown gel (BG, a basal gel that may be dark green and is common in ochre formations) or a red cloudy (RC), brown ring (BR) that terminates (dominated aerobically) with a brown cloud (BC). All these reactions involve bioaccumulation of ferric iron.

Reductive conditions may be recognized by a green cloudy (GC) reaction in which ferrous iron forms dominate. The terminal reaction of active mixed bacterial flora in a consorm is the generation of a black liquid (BL). In IRB BART, the sequence of the reactions dictates the reaction pattern signature (RPS) used to define the specific IRB within a consorm.

Incubation temperatures for BART methods are covered in Section 9.3. Common practice for surface biosphere samples is to use room temperature; the ideal should be within $4°$ of the natural environment from which the sample was taken. Clearly the temperature selected for sample incubation affects rates of activity and growth. TLs depend greatly on the appropriate incubation temperature.

Each BART indicates a second line (TL) range over which positive detections of bacterial activity may be observed. Time lapse is defined as the time from the start of the test to the first observation of the reaction. Precision depends on the frequency of observation. Fully quantitative precision can be achieved with the visual BART reader system (VBR72) that takes digital images on a preset schedule. TLs for very active samples may be registered in seconds; the normal intervals between image sequences may range from 1 minute for samples with very high levels of activity and 15 minutes for samples with low levels. The shorter the TL, the greater the activity level and the larger the active bacterial population in a sample. Populations are measured as predicted active cells per milliliter or per gram (pac/ml or pac/g) directly reflect activity in the BART. The VBR72 obtains population data as soon as the reaction is observed by scanning the time lapses for reactions. The software is already set to interpret TLs generated at 22°C and 28°C.

Tables 10.1 through 10.16 provide characterization data. The BART results shown list three possible alternative reaction patterns and a single TL based on whether a range has been defined, is not detectable (ND), or is not known (NK). A reaction and TL may appear or no detection may occur. If the occurrence of a reaction or TL is more likely than no detection, the ND notation will be bracketed after the reaction or TL value, i.e., the reaction *could* occur on occasion. A TL number (in days) followed by a dash and ND indicates a positive reaction generating a time lapse no shorter than the number shown. If no ND with or without brackets is shown, a TL would normally be generated by that type of consorm. All data in the tables is subject to the ability of a sample to cause a reaction. As a consorm matures, its activity level is expected to decline. Such declines may be manifestly affected by post-sampling handling of the sample before the start of the testing.

The challenge for a technologist in the field collecting samples for BART who wants to start testing after he returns to the laboratory is how to maintain the samples until testing can be performed. Keeping samples more than a day creates problems because the bacteria in the sample will change activity levels and dominant communities may shift. To "level the playing field," all of the samples should follow a standard protocol and storage time should be reasonably consistent. Most bacteria enter a dormant state when temperatures drop below 7°C. This simply requires a refrigerated environment (4 ± 2°C)—a small portable refrigerator or ice-packed cooler. Sample bottles should not be packed too tightly because that will create temperature variations.

Impacts on the bacterial activity of a sample depend on the original temperature of the sample. At storage temperatures of 4 ± 2°C, most bacteria become far less active and the samples may be stored generally up to three weeks before some of the bacteria become active again. Whether a sample is kept for 1 day or 3 weeks, the bacteria within the same would achieve a common level of inactivity (static state). For this reason, samples stored 1 day cannot be compared with samples stored for 3 weeks.

In setting up BART using these "aged" samples, it is very important to allow the samples to return to room temperature. Simply place the sample containers on a bench without stacking them or pushing them together. A good flow of air around each bottle ensures that all samples will reach room temperature (22 ± 2°C) before the onset of testing. Of course, cooling followed by warming impacts all samples and levels of bacterial activity, hopefully in a consistent manner. TLs may lengthen because of he additional time required for the bacteria to adapt.

While reactions may not be affected by prolonged storage, the TL (and hence the prediction of predicted active cells per milliliter or gram) lengthens with smaller predicted populations. However, the data generated may be used comparatively for samples subjected to the same storage regimen. In summary, storing samples in a refrigerator up to 3 weeks means that BART may be performed for the major bacterial communities but may show lengthening of TL and changes in the reaction pattern signature.

Sample sizes required for BART depend on the origin of the consorm. Clear or relatively clear water samples may be applied directly as 15 ml aliquots to start the tester. Turbid water samples and solid or semi-solid materials require some preparation to ensure that the tester does not become opaque at the outset of the test and thus obscure the reactions. Turbid water samples can be diluted 10-fold to improve clarity and allow testing. When a population is calculated using VBR72 or BART SOFT then the population would have to be multiplied by ten to arrive at the correct population.

For solid and semisolid samples, BART requires a protocol that would allow direct admission of the samples. Samples of 0.1 or 0.5 g are recommended; the smaller amount is applied to samples likely to generate dispersive turbidity in the final sample, for example, a clay loam or dense black sedimentary slime. The BART inner cap is removed, turned upside down. and the sterile BART ball rolled into the sterile side of the upturned cap. The 0.1 or 0.5 g sample is added to the tester (inner vial) without significant smearing of the vial walls; then 15 ml of sterile distilled water is added to start the test. Deionized water is not recommended because it is not sterile and may contain bacteria. After the water is added, the BART ball is rolled into the vial and the cap screwed onto the vial. This technique prevents particulate material from the sample to lodge around the BART ball and prevent it from floating correctly.

Testing for APB or HAB requires an additional step: indicator chemicals dried inside the cap must be dissolved and released when the water has been added and the ball is rolled into the tester. For both tests, 0.1ml water is added to the inside of the cap after the BART ball is returned to the vial. Leave the water in the cap for 30 seconds, tip the contents into the tester and screw the cap down. If the tester contains floating particulates, allow the indicator chemicals to diffuse into the water. If no floaters are apparent, the tester can be gently shaken three times without turning the tester over (causing particles to lodge around the ball and yield imprecise results).

When dilutions are required to prepare the tester, the predicted population projection must be adjusted. Because of dilution, the final population number will be lower. Correction is achieved by multiplying the projected population by a factor of 10 for 1.5 ml/g, 30 for 0.5 ml/g, and 150 for 0.1 ml/g. Tables 10.1 to 10.16 assume the correction factors were applied to all samples except clear or mildly turbid waters.

11 Defining Bacterial Consorms in Gridded Atlas Format

11.1 INTRODUCTION

Reductionism dictates that science must define everything in its simplest form differentiated from factors that may interfere with the perceived precision created by this approach. This gridded atlas utilizing oxidation–reduction potential [ORP, expressed as factorial millivolts (fmv)] and water viscosity (fcP) characteristics employs holistic approaches based upon the general functioning of a bacterial consorm. The objective is to define the functioning of the whole consorm indicated by its form and function. This definition method does not depend on any single component bacterial strain, species, or genus. It identifies a consorm based on its capacity to adapt biochemically, physiologically, and structurally to retain its efficient function even if several of the initiating bacterial strains are eliminated. Such changes mainly reflect changes of the environment in which a consorm functions. This chapter examines the interactions of consorms to each other and to the localized environment within which they thrive.

The 13 figures relate to the positioning of some of the bacterial consorms defined by testing. Figure 11.1 defines the relationships of fmv and fcP with the basic forms of bacterial consorms. The role of the static water table in porous and fractured media is defined along with the impacts of ORP and water viscosity. Animal and plant habitats are limited to the oxidative surface biosphere; plants are restricted to the photic zone and animals to environments with available oxygen. Habitat limitations for animals are predefined by their needs for oxygen and suitable organic food sources (Figure 11.2). The diurnal need of plants for photosynthesis restricts their activity to regions where light can penetrate air- or water-filled liquid, fractured, or porous environments. Figure 11.3 illustrates these constraints.

Microorganisms reveal broader spectra of activity (illustrated via fmv:fcP grids) than plants and animals. The domain of prokaryotic bacteria is very large and includes very reductive regimes in which animals and plants could not survive. These bacteria are also active in very viscid waters (see Figure 11.4). Eukaryotic microorganisms lack the environmental flexibility of prokaryotic types and are associated with limitations of the surface biosphere (Figure 11.5). This greater environmental flexibility of prokaryotic bacteria is reflected in their dominant roles in biomasses found in reductive regions of plants (e.g., deeper saturated geologic media).

Five figures deal with specific relationships between consormial groups and various habitats linked to animal or plant activities in various ways. Figure 11.6 defines

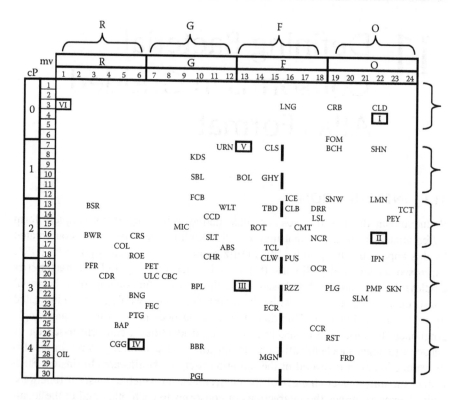

FIGURE 11.1 Basic patterns in fmv:fcP grid. The divisions of the four-sectioned x axis grid are determined by whether conditions are reductive (R), gradient (G), located the oxidative–reductive interface front (F), or oxidative (O). The y axis has five sections based on water viscosity: 0 = water dispersed into air, 1 = water in a free liquid state, 2, 3, and 4 are log increases in the fcP indicating increased viscosity. A discontinuous vertical line at the border indicates fmv movement from reductive (left) to oxidative (right).

the bacterial consorms (primarily 2, 16-21 RZZ) associated with plant activities. Major bacterial activity occurs in the biochemicals streaming through the intestines of all animals and varies among different groups of animals. Figure 11.7 indicates the bacterial consorms associated with intestinal streaming in non-herbivorous mammals. Since foods are raw or processed products from animals and plants, Figure 11.8 shows the principal consorms that can spoil foods; spoilage is defined as an event involving bacterial consorms whereby a product becomes esthetically, hygienically, or structurally unacceptable.

Certain bacterial consorms (pathogens) are also connected to disease-causing activities. Traditional recognition of disease involved recognition of a pathogen as a single culturable species using Koch's postulates. Failure to culture meant that: (1) the pathogen could not be cultured and isolated; or (2) the pathogenic function resulted from consorm containing more than a single pathogenic strain. Both possibilities led to the definitions of many cancers as cellular dysfunctions caused by consorms or unculturable bacterial strains. Figure 11.9 depicts non-enteric pathogens of mammals;

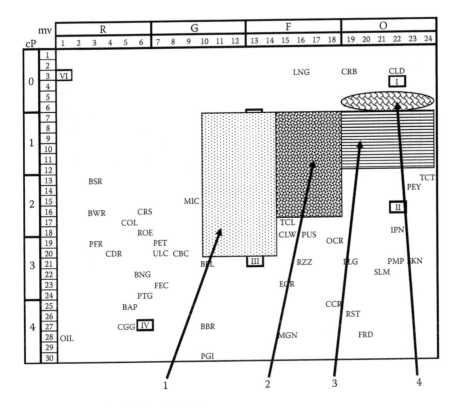

FIGURE 11.2 Animal habitat limitations.

the diseases may include consormial activities not recognized by the reductionist approaches to disease diagnosis. Potable surface waters and groundwaters are natural vehicles that can spread pathogens. Figure 11.10 defines the bacterial consorms linked potential hygiene risks in waters.

One largely ignored but major function of geological media is the recycling of organic carbon from the surface biomass into hydrocarbons (where inorganic elements are stripped from the molecules) followed by biodegradation of the hydrocarbons when they reascend into the oxidative regions of the biosphere. The reductive (primary descending) phase of carbon recycling is shown in Figure 11.11. Figure 11.12 illustrates oxidative (secondary ascending) phase. This recycling of carbon from complex organic forms in oxidative regions to hydrocarbonaceous forms under reductive conditions is simply another natural cycle of an element important to life. Failures of gas and oil wells to deliver acceptable fractions of reduced carbon reserves may be in part due to the activities of bacteria that plug the well environments, thus preventing flows and ultimately causing abandonments of wells. Extraction and injection water wells are frequently subject to flow interception resulting in plugging and production failures. Figure 11.13 illustrates the principal bacterial consorms that generate plugging biomasses that limit production and ultimately cause total failure. These step-like failures from plugging and similar

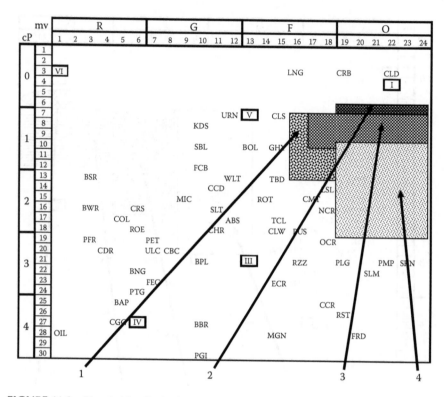

FIGURE 11.3   Plant habitat limitations.

functional failures also occur in gas and oil wells but are not widely recognized as results of active biomass plugging.

11.2   BASIC FMV: FCP GRID

Figure 11.1 shows 20 major grids in Figure 11.1. Four differentiate fmv horizontally and five differentiate fcP vertically. Each major grid supports a different type of consorm requiring specific oxidative–reductive conditions and water viscosities. A critical factor in defining boundaries of consorms is whether the water saturates a geologic medium (commonly porous) at the oxidative–reductive interface. This characteristic is shown as a discontinuous vertical line and defines the region where biomass tends to congregate.

11.3   LIMITATIONS OF ANIMAL HABITATS ON GRIDDED ATLAS

Figure 11.2 indicates the four common atlas regions where animals are metabolically active. The most reductive environments for activities are the deeper confines of the body (soma) where significant stressors are applied via restriction of oxygen supply (1). Around the oxidative–reductive interface (F), the environment contains enough oxygen for animals such as insects and worms to live at and above the static

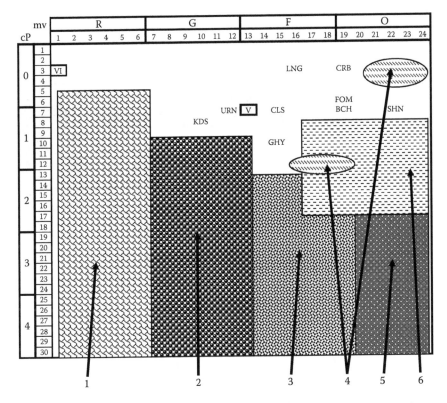

FIGURE 11.4 Dominant prokaryotic consormial domains. Alpha five and alpha six consorms are not included as dominant organisms because their role is speculative (alpha six) or subject to poor definition (alpha five). Alpha five and alpha six probably include the forms that fit into the alpha two and alpha three consorms in the figure.

water line. Burrowing animals must assure their air supplies by constructing tunnels (2). Animals within a water column (3) differentiate into layers, with zooplankton near the surface. Fish and water-borne mammals occupying most of the oxygenated water column. Insects and worms are bottom feeders that penetrate the surfaces of muds. Many other animals and birds occupy the land surface biosphere (4). Animals are primary or secondary consumers of generating plant biomasses and depend on oxygen to drive respiratory function.

11.4 LIMITATIONS OF PLANT HABITATS ON GRIDDED ATLAS

Plants generally grow in photic zones where transmitted or sorbed sunlight can penetrate their habitats. Figure 11.3 therefore displays the areas in the gridded atlas where light penetrates and water is sufficient to allow growth. The four major plant groups on the atlas are: (1) plants that grow above ground, particularly in warm, moist climates such as rain forests, (2) the temperate plants and aquatic weeds that grow attached to soil or suspended in waters, usually at the air:water interface,

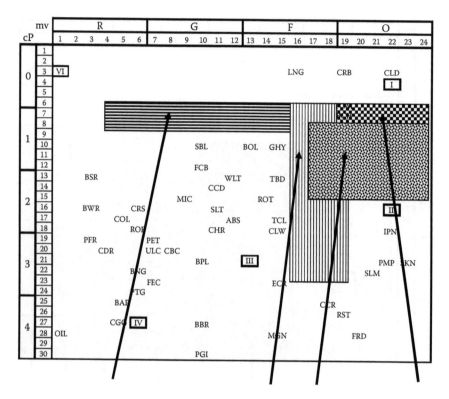

FIGURE 11.5 Dominant microbiological eukaryotic domains.

(3) suspended plants totally immersed in liquid water, growing within the limits imposed by the photic zone, and (4) algae growing in porous soils that light can penetrate.

Total plant biomass includes two prime components: photosynthesizing organs (leaves) exposed to light during the day and attachment orga (roots) that provide support and procure water and nutrients from the underpinning soil and water.

Only the soil-based algae lack organs of attachment (1) and grow in soils above the static water level where light can penetrate their depths. Large rain forest (2) biomasses are composed of photosynthesizing organs set well above ground to allow efficient carbon dioxide capture. The large biomass supports the organs bring water and nutrients from the roots. A small part of the biomass is dedicated to the photosynthesizing organs; most of the biomass is required to develop and maintain the supporting structures. After a long maturation period, rain forest tree trunks provide economically valuable woods. Surface dwelling plants (3) tend to be seasonal, generating annual growths with seeds for propagation or are perennial growths that harden to survive inclement winter. The fourth group lives in water and sediments (4) in the photic zone. They may float free or attach to underpinning sediment. Water and nutrients are absorbed more directly by the biomass; attachments ensure that the plants stay in favorable environments.

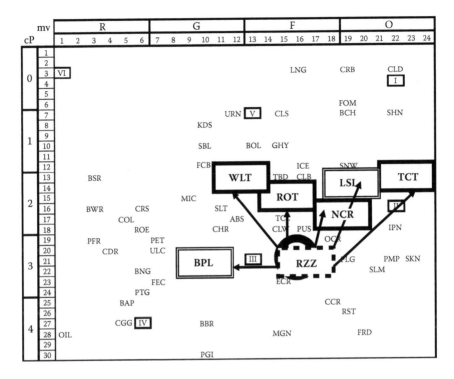

FIGURE 11.6   Bacterial consorms associated with plant activities.

11.5   DOMINANT PROKARYOTIC CONSORMIAL DOMAINS

Prokaryotic bacterial consorms span the widest range of fmv and fcP in the atlas grid; the six major groupings are shown in Figure 11.4. Consorms are grouped in the atlas by position. Four groups roughly follow the four major fmv conditions (R, G, F, and O) and extend through various viscosities of water (values of 1 or more), with greater differentiation in the oxidative (O) region. Differentiation into six groups is based on the nature of the substrate (fractured or porous geological media). These consorms: (1) grow primarily in deep reductive fractured or porous geologic zones and aiding the generation of hydrocarbons from descending organic materials; (2) inhabit the reductive interfacial zone, often in porous media rich in nutrients to support anaerobic activities; (3) occupy porous media, usually soils that perch water to create a static water level; (4) form around solid crystallized water as ice or remain suspended in the air as bionucleates; (5) live in viscid liquid waters containing high biocolloidal and suspended particulate levels; and (6) are active in fairly free liquid water where they can function without high biocolloidal contents.

11.6   DOMINANT MICROBIOLOGICAL EUKARYOTIC DOMAINS

Figure 11.5 illustrates the relationships of eukaryotic microorganisms that include protozoa, microalgae, fungi, and anaerobic photosynthesizing prokaryotic bacteria.

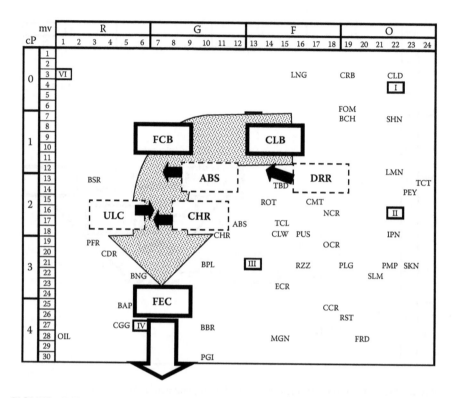

FIGURE 11.7 Bacterial consorms associated with non-herbivoral intestinal streaming. Grey curved arrow defines passage route through intestine. Three black boxes show normal bacterial consorms of intestine. Four dashed boxes with black arrows indicate major pathogenic consorms. Open arrow indicates discharge of feces.

Generally the eukaryotic microorganisms are found in the less viscid waters where they have access to oxygen. Exceptions are the anaerobic photosynthesizing prokaryotic bacteria (1) that undertake photosynthetic functions without generating oxygen except under reductive conditions, like microalgae. They require sunlight but at longer wavelengths and are found in very reductive eutrophic conditions such as in sanitary wastewater lagoons. The waters may turn from growth of these bacteria.

Fungi (2) dominate in environments within porous or fractured media, where the oxidative–reductive interface is present but water does not saturate. Bacterial consorms and fungi constantly compete, and the fungi dominate under low moisture conditions (<18%). Bacteria dominate completely at moisture >80%. The transitional zone (>19 to <79% water) is where the active competition takes place. Many bacteria can adapt to restricted oxygen availability and this ability is a critical factor in the competition. Protozoa (3) are less tolerant to porous media than fungi; they dominate in liquid waters, particularly those containing suspended particulates. Protozoans also lack tolerance of low-oxygen concentrations. They are oxidative degraders of ingested organics, creating high-oxygen demand and low efficiencies in synthetic functions. Microalgae (4), like all members of the plant kingdom, photosynthesize

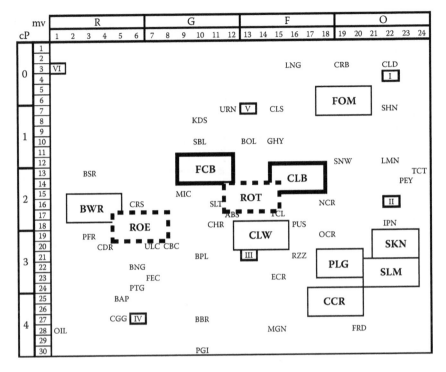

FIGURE 11.8 Bacterial consorms involved in food spoilage. Hygiene risk consorms are shown in thick lined boxes, extrinsic consormial consorms in dashed boxes, and intrinsic consorms in thin walled boxes.

via the production of oxygen and organic products. At night, however, they revert to respiratory functions that demand oxygen. Microalgae are therefore limited to photic zones in waters and oxidative soils but can compete with heterotrophic bacteria for organics in the absence of light.

11.7 BACTERIAL CONSORMS ASSOCIATED WITH PLANT ACTIVITIES

Most interactions of plants and bacteria occur in the saturated and semi-saturated porous media of soil into which the roots penetrate. The roots gain access to water and nutrients and also provide a level of stability. The plant–soil interfaces provide natural junctures for bacterial consorms to operate within the rhizosphere as 2, 16-21 RZZ. A mutually beneficial barter system exists between plants and RZZ. The bacterial consorm gains from the movement of synthesized organic and oxygen through the roots and into the soil. The plant roots can extract water and mineral nutrients from soil. The RZZ consorm first extracts minerals and water from soil voids and holds them within extracellular polymeric substances (EPS). The EPS provide a secondary benefit: stimulants, penetrants, and binders that maintain the structural integrity of the soil.

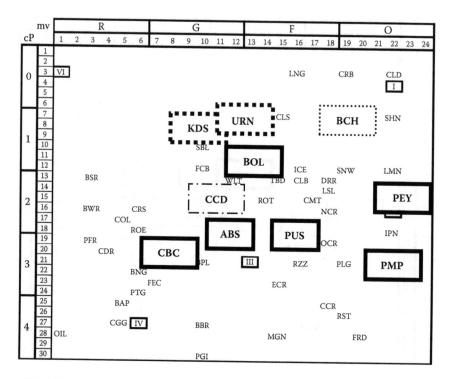

FIGURE 11.9 Mammalian consormial pathogens in body (excluding intestine). Bacterial consorms possibly linked to diseases in mammalian organs are: urinary tract (dashed box), lungs (dotted box), teeth (dot–dash box), and skin and tissues (thick lined box).

Figure 11-6 illustrates the role of 2, 16-21 RZZ as the vehicle of passage between the root systems and the bacterial consorms that can become parasitic and/or pathogenic to plants by causing dysfunctions that lead to functional failures.

Stresses can involve specific site symptoms or lead to generalized plant health failure. Consorm 2, 16-21 RZZ (dashed box) appears when a plant is healthy but other bacterial consorms may cause stresses of a direct (thick lined box) or indirect (double thin lined box) nature. Indirect stresses involve the development of an impermeable barrier through which water cannot penetrate to reach the roots and RZZ. Tight impermeable barriers form at the oxidative–reductive interface, often at the static water level to form a reductive black plug layer (3, 10-21 BPL) that prevents water movement and directly competes with plant roots and causes dieback. Looser layers of slime (resembling a sandwich) form when 3, 18-14 LSL generates lateral slimes that tend to divert the water to the side.

Four bacterial consorms can create direct plant stresses that can lead to death. They attack at different plant sites. Three attack plant tissues in different ways. Generalized destruction by tissue collapse of tissues (necrosis) is caused by 2, 18-16 NCR. If tissues lose strength and partially collapse, the cause is 2, 14-15 ROT, generalized rotting. If the tissues are penetrated and infestation materializes only in the phloem and xylem that handle nutrients, water flows, and turgor pressures,

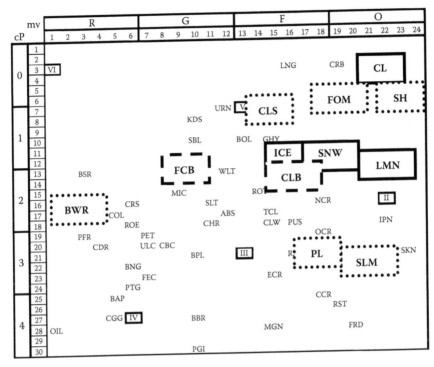

FIGURE 11.10 Consorms associated with water quality issues. Water quality involves normal bacterial consorms found in the various states of water (thick lined box); water quality that has become challenged (dotted boxes); and health risk consorms are shown as dashed boxes.

collapse will occur around the point of infestation. This is caused 2, 12-13 WLT (WILT). Some bacteria infest plant canopies. EPS binds the leaves, gumming the foliage into a set position. Rain or irrigation water becomes trapped in the EPS and does not precipitate downward into the soil. This type of effect is known as thatch (2, 24-13 TCT. Grass located beneath a thatch will dry out and die, leaving denuded soil where the thatch occurred.

11.8 BACTERIAL CONSORMS ASSOCIATED WITH NON-HERBIVORAL INTESTINAL STREAMING

In more evolved animal species, mature animals receive solid and liquid food by ingesting and digesting them; the digested products are being circulated into the body organs as needed. The site for digestion is the intestine and the phases of digestion are containment, mechanical and chemical degradation, extraction, and ejection. The intestine provides for the growth of several bacterial consorms that may then assist with the process or infest the site (Figure 11.7).

The intestine mechanism allows food and drink taken into the body (soma) to be digested. It is perhaps significant that we enjoy a delightful and nutritious meal, then

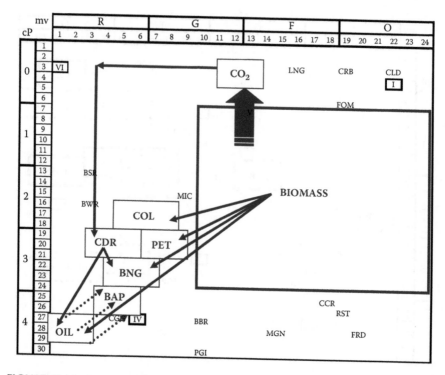

FIGURE 11.11 Bacterial consorms involved in reduction of descending organics to hydrocarbonaceous or carbon-rich products. Biomass is generated (thick lined box at right) with descending carbon moving (arrows) toward very reductive conditions deep in the geologic crust. Carbon moves into the reductive regime via two routes. First, the biomass may become respired or combusted via the release of carbon dioxide (vertical black arrow); under reductive conditions, the gas may be reduced. Second, the growing biomass may descend into the crust and be reductively converted (solid black arrows) into reduced carbon that forms asphaltenes (dotted arrows).

surrender it to the digestive chemicals and bacterial consorms. Little is known of the bacteriology of the intestine; the focus to date has been on the complex biochemistry of the digestive reactions.

Three consorms are active in digestion: coliform bacteria (2, 16-13 CLB), fecal coliform bacteria (2, 10-12 FCB), and the little known consorm that handles the recovery of water and final nutrients, leaving a dense bacteriologically rich fecal pellet to be ejected. The consorm responsible for this conservation program (removing water and nutrients) is 2, 07-23 FEC; the bacteria are ejected in the fecal pellets.

Four pathogenic bacterial consorms can infest animal bodies: 2, 06-19 ULC causes ulcers; 2, 11-18 CHR causes cholera; 2, 12-17 ABS causes abscesses; and 2, 18-13 DRR causes diarrhea and functional intestinal failures. In each case, a specific pathogenic strain may cause the disease, but the symptoms arise from the consormial activity driving the pathogen into a dominant role.

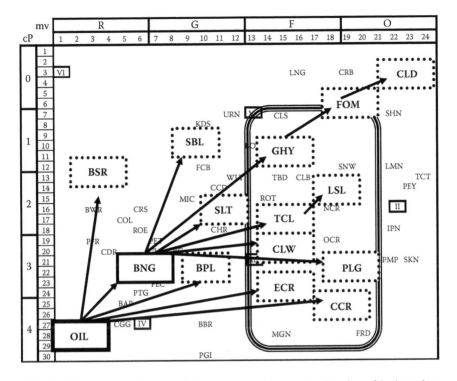

FIGURE 11.12 Bacterial consormial interceptors in upward migration of hydrocarbons. The major oxidative–reductive interface is shown as a triple bordered box. Solid arrows indicate common route to oxidative reassimilation of hydrocarbon fractions generated by the consorms (thick lined boxes) toward the various consorms that accumulate and assimilate carbon back into the biomass.

11.9 BACTERIAL CONSORMS INVOLVED IN SPOILAGE OF FOODS

Foods are very vulnerable to bacterial spoilage because their nutritional value is generated by some blend of proteins, carbohydrates, and fats within a water-based medium. As the water of a food climbs from 18 to 85%, it exhibits increasing potential for spoilage through bacterial consormial activity. The food industry attempts to suppress the potential for bacteriologically influenced spoilage by creating antagonistic environments (e.g., reducing pH, adding biocides) or sterilizing packages to creates durable abiotic zones around the food to prevent bacteriologically influenced spoilage. Spoilage automatically represents a cost to a food packager and the cost is passed along the production line to the consumer. Urbanization led to the need to effectively preserve foods and drinks. Appert devised the first preservation technique involving heat in the early 19th century. Pasteur's work a half century later led to pasteurization of milk and beers.

Consumer acceptance is a major economic factor and packaging plays a major role in acceptance. Consumers prefer to see foodstuffs directly before they make

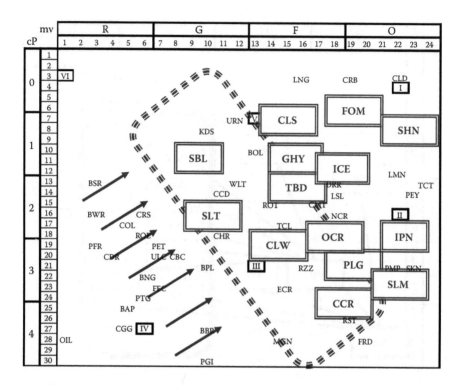

FIGURE 11.13 Consormial interceptors of groundwater flows that cause plugging in porous and fractured media. Groundwater flow from reductive regions is shown by arrows; main region of oxidative–reductive interface as rectangular dashed triple line; bacterial consorms as double bordered box.

purchases to assure themselves no spoilage is present. If a consumer must open a food (e.g., canned meat) to see whether it is spoiled, he has a natural tendency to look at a product carefully and even smell it. Food spoilage has been subjected to reductionist analysis for about 200 years, but only rarely has it been successful because spoilage is commonly a consormial affair involving many different communities of microorganisms. Figure 11.8 shows in gridded atlas format the range of bacterial consorms that cause hygiene risks along with intrinsic and extrinsic challenges to food longevity.

Hygiene risks in foodstuffs are attributed to inhabitants of mammalian intestines that present the potential for foodstuffs to carry pathogenic strains of bacteria that may trigger disease. Two consorms subject animals to hygiene risk (2, 16-13 CLB and 2, 10-12 FCB). Here the former are enteric bacteria normally found in intestinal tract; while better defined latter relates to *Escherichia coli*.

Extrinsic spoilage arises from the entrance of a consorm directly into a foodstuff. One example is the extrinsic contamination of the outer shells of eggs followed by penetration of the protective shell, membrane, and lysozyme to cause spoilage (2, 06-18 ROE). A second example is extrinsic spoilage in which a consorm (2, 14-15 ROT) penetrates plant tissues through their waxy cuticles and causes generalized decay of tissue structures into a pulpy mess.

Intrinsic spoilage occurs when the infesting consorm is already present in a foodstuff that has not been sterilized or is subjected to post-sterilization contamination. Clear beverages can be compromised by cloudy growths (2, 15-18 CLW), colloidal turbidity that generates more viscosity (2, 15-13 TBD) or causes the liquid to blacken (3, 03-16 BWR). Food or water can spoil at the interface between packaging and product by the generation of slimes (2, 21-22 SLM) or the formation of skin (2, 23-21 SKN). Under some conditions, consorms may generate foams as gases (1, 19-06 FOM) that render a foodstuff unacceptable. If a food has a porous matrix, bacterial consorms can infest its and create biomass plugs (2, 19-21 PLG) that slightly changes the texture of the product. Foods with higher inorganic contents may be subjected to spoilage by hardening crust-like growths (3, 18-25 CCR) that render them unacceptable.

11.10 MAMMALIAN CONSORMIAL NON-ENTERIC PATHOGENS ON GRIDDED ATLAS

Traditionally, bacterially perpetrated diseases were considered to originate from a single strain that disrupted the normal physiological functioning of the impacted host. This premise has two major flaws: (1) it assumes that the single strain isolated is the sole source of the disease (confirmed by the application of Koch's postulates) and by default assumes that a disease is not bacterial in origin if a single strain cannot be isolated; and (2) the premise that the disease may have resulted from infestation by more than one strain (community or consorm) or a failure in normal functioning of a bacterial consorms already active in the body is not considered. These shortcomings in disease diagnosis significantly affect human health because the cause is not isolated (due to failure to isolate and culture the strain or recognize a consormial infestation).

Figure 11.9 illustrates that bacterial consorms may be significant in mammalian bodies (excluding consorms of the intestinal tract). Identifying technologies are only now becoming established (see Chapters 8 and 9). The gridded atlas indicates which bacterial consorms may act in different parts of the body.

This premise of bacterial consorm activity casts a wider net to determine whether certain diseases are bacteriologically influenced (by strain or consorm). Certain cancers may be unrecognized bacteriologically triggered events. The RASI-MIDI S43720 technique applied to cancerous tissues may generate a FAME chromatograph indicating the consistent presence of bacterial consorms. Urinary tract infections have two major sites of infestation. First, infection of the tract may culminate in infection of the bladder involving many strains of bacteria within an infesting consorm (2, 12-07 URN). Second, in the kidneys, bacteriologically influenced concretions (2, 10-08 KDS) may form (kidney) stones that can plug the tract and cause severe discomfort.

Teeth are very resistant to infection and decay. They contain coatings of enamel over a dentine base. Teeth diseases usually involve direct corrosion of the enamel extending into the dentine or infection at or below the gum line. Carietic conditions are generally reductive and produce fermentative acid daughter products that erode the enamel and dentine. 2, 11-14 CCD is the consorm that thrives under these reductive conditions, protected from the oxidative environment in the blood and the various body defense mechanisms.

Lung tissues are very vulnerable to infection because the lungs handle the body's oxygen intake under conditions nutritionally favorable for invading bacterial consorms. A number of defense mechanisms such as filtration of the air and the removal by white blood cells of alien particles from the lung tissues protect the lungs. A sustained lung infection usually starts when bronchitis (2, 19-07 BCH) consorm gains a flagella hold on lung tissues. Bronchial infections can become chronic if the body's defenses are too weak to contain the infestation. Sometimes conditions can become suitable for the onset of tuberculosis triggered by *Mycobacterium tuberculosum.*

The eyes are the organs that are most vulnerable to infestation but they rarely become infected. Eyes are directly exposed to the outside environment but the body defense systems usually prevent major infections. Defense systems that protect the eyes center on three components: (1) the crystalline lens and cornea act as blockers to infection; (2) constant applications of lysozymes destroy bacterial cells; and (3) constant washing action removes particulate materials from the eyes. Infections around the eye balls can cause the eyes and surrounding tissues to turn pink; the condition is known as pink eye. Consorm 2, 23-14 PEY is associated with these infestations.

Skin and underpinning muscle tissues are vulnerable to consormial invasion within a number of the strata. Small infestations of the skin by 2, 22-21 PMP can cause pimples. Skin infestations deeper in tissues can extend laterally and press the skin tissues upward into a dome. This condition is a boil associated with 2, 13-10 BOL. If the infection site is deeper, an abscess (2, 12-17 ABS) or carbuncle (2, 08-20 CBC) may ensue and the pressurized tissues may break apart, releasing pus (2, 18-18 PUS).

11.11   BACTERIAL CONSORMS ASSOCIATED WITH WATER QUALITY ISSUES

Figure 11-10 illustrates bacterial consorms that are active in water and impact quality. Consumer water quality issues are high clarity, absence of odor, and certification of the absence of coliform bacteria. In Nature, water serves as the common substrate and solvent for all life forms. In the water cycle, clouds form the upper stratum of water that is bionucleated in drops by EPS generated by consorm 1, 22-03 CLD. When the temperature in the clouds drops to freezing range, EPS cluster crystallized water ice into solid forms (1, 19-12 SNO) that float to the ground. If the water pools and the temperature continues to drop, the freezing water will form ice (1, 16-12 ICE). When bacterial consorms are active, they can produce ice crystals above the normal freezing point of water. If water descends into a deep (oceanic) body, it will become more saline at greater depths. Beneath the photic zone, a layer of biomass known as the deep scattering layer is where intense microbiological activities take place. Commonly between 400 and 1,000 meter depths, bioluminescent bacteria (2, 22-12 LMN) produce a layer that glows blue in the dark.

Water quality can also be affected by the growth of bacterial consorms. Water may become cloudy (2, 15-07 CLS) under oxidative conditions or blackened under reductive conditions (3. 03-16 BWR). High levels of organics can cause water to develop foam at the water:air interface (1, 19-06 FOM) if the water is turbulent or become sheen-like (2, 22-07 SHN) if the water surface is static. If the water has a high organic content and remains oxidative and not turbulent, slimes can form in the water (2, 21-22 SLM). In groundwaters rich in iron, foci of ferric-rich bacteria (2, 10-12 FCB) can develop at the oxidative–reductive interface, particularly if it coincides with the static water level. In a porous or fractured geological medium with some groundwater flow, biomass plugs (2, 19-21 PLG) can restrict or stop groundwater flow.

11.12 BACTERIAL CONSORMS INVOLVED IN OIL, GAS, AND COAL PRODUCTION IN GEOLOGICAL MEDIA

In addition to Earth's water cycle, a carbon cycle essentially supports life and certain life forms generate energy reserves in the form of natural gas, oil, and coals. This cycle starts under photo-oxidative conditions where energy derived from the sun is trapped during generation of a biomass based on photosynthesis. The biomass is recycled in the surface biosphere under oxidative conditions through light–darkness diurnal cycles by animals that feed on the plant biomass and animals that feed the biomass. Underpinning this are bacterial biomasses functioning in soils, waters, and in the intestines of animals. The result is a relatively inefficient re-use of the organic carbon biomass.

Via gravity, some of the dispersing, degrading biomass moves downward into the reductive water-saturated regions of the Earth's crust. Dominant anaerobic bacteria strip the dispersing organics of essential elements such as oxygen, nitrogen, phosphorus, sulfur, potassium, iron, and a variety of trace nutrients. Stripping involves movement of elements from degrading organics to the incumbent biomass in the reductive water-saturated crust. Within the oxidative regime, organic carbon is respired or combusted and produces carbon dioxide. Progressive removal of these elements in the reductive lateral zones continues until only hydrocarbons or elemental carbon remain. The deep geologic pools of carbon are dominated by crude oils as petroleum hydrocarbons, natural gases, and coals, thus completing the descending phase of the carbon cycle illustrated in Figure 11.11.

If carbon dioxide enters the crust deliberately via injection or passive permeation, the reduction conditions will cause the stripping of the oxygen from the carbon dioxide and concomitant generation of hydrocarbons. The bacterial consorm that undertakes these reductive activities is 4, 04-20 CDR and the final product may be biogenic natural gas through 4, 06-22 BNG or petroleum hydrocarbons through 4, 01-28 OIL. While some of the carbon feeding these highly reductive activities comes from carbon dioxide, most is derived from the covert reduction of the descending biomass over four possible routes on the atlas. The organics may be reduced to elemental carbon (4, 05-17 COL), involve a less complete reduction with a higher residual organic content as peat (4, 07-19 PET), or undergo a direct biogenic reduction to natural gases (primarily methane) by 4, 06-22 BNG and to oil (4, 01-28 OIL). If oil has significant water content. some of the water may be bacteriologically "mined" by 4, 05-25 BAP consorms from the oils. This water is retained in asphaltenes as a thick black "goop" that plugs wells and pipelines. What goes down must come up to compete the carbon cycle; this is addressed in the next section.

11.13   BACTERIAL CONSORMIAL INTERCEPTORS IN UPWARD MIGRATION OF HYDROCARBONS

The descending carbon cycle component in the crust leading to the bacteriologically influenced reduction of organic carbon to hydrocarbons is discussed in Section 11.12. In the ascending component of the carbon cycle, hydrocarbons return to the oxidative surface biosphere to complete the cycle and generate renewable oil and natural gas resources. Two mechanisms comprise the ascending cycle. The first is the deliberate recovery of the hydrocarbons using extraction wells or mining techniques; the second is the natural upward migration of the lighter hydrocarbon fractions as natural gases seeping through the crust or volatile daughter products that move upward along a diffusion gradient toward the oxidative surface biosphere. While deliberate extraction is used as a primary energy driver, with combustion converting most hydrocarbons to carbon dioxide, the natural diffusive processes are likely to lead to the assimilation of these reduced forms of carbon into the oxidative surface biosphere. Figure 11.12 illustrates the consorms that are active in the reintegration of hydrocarbons into the biomass.

Hydrocarbons feeding the growth of the biomass incorporate natural gases seeping up through the crust along with biogenic methane produced on the reductive side of the oxidative–reductive interface. Light fractions of oils that are volatile or highly mobile in groundwater are also likely to be assimilated into the biomass.

Consorms associated with the reductive generation of petroleum hydrocarbons (4, 01-28 OIL) and the biogenic generation of natural gases (4, 06-22 BNG) are the prime generators of basic forms of reduced carbon that can be recycled when natural diffusion into the oxidative zones in the biosphere occurs. Figure 11.12 assumes that these forms of carbon will ascend via the groundwater or seep into the surface biosphere over time.

Radical seepages may occur through deep-ocean black smokers in which the bacterial consorm 4, 03-13 BSR interacts with the reductive natural gas-rich hot waters moving up through vents into the oxidative cooler oceanic waters. Seepages of natural gas may also be contained by gas hydrates as bacteriologically influenced ice crystals form and become saturated with methane gas in part through the activities of 5, 15-10 GHY. This gas–ice hydrate is stable up to 7°C and generates freezing properties in the EPS generated by the consorm. Methane is held within a stabilized reductive structure until oxidative, electrolytic, or ice melting causes the methane to be released and degrade or be assimilated into the consormial biomass.

Natural gases and the volatile and/or diffusible petroleum hydrocarbons travel through groundwater along diffusion gradients, moving upward from higher to lower concentrations until they reach the oxidative–reductive interface. A number of habitats can generate conditions along the interface formed by the dominant bacteriological biomass that forms the primary barrier preventing escape of these products into the atmosphere. Several consorms are capable of degrading or assimilating the volatile and gaseous hydrocarbons in the first barrier dictated by the position of the oxidative–reductive interface. Consorms in the plethora of bacteria that can degrade or assimilate hydrocarbons ascending from the reductive crust regions crust include 3, 10-21 BPL that forms reductive plugs and restricts movement prior to oxidative degradation; 3, 10-10 SBL that forms in groundwater conduits in porous media, diverting waters to the line of least resistance (commonly upward); and 3, 18-14 LSL that holds reductive waters rich in hydrocarbons, triggering their degradation and assimilation within the upper more oxidative layers. If the hydrocarbons enter the root zones of plants, it is highly probable that 2, 16-21 will intercede; assimilation and degradation can also occur in concretions (3, 18-25 CCR), tubercles (3, 15-17 TCL), silts (2, 11-16 SLT), and plugs (2, 19-21 PLG) that restrict movement and control assimilation of hydrocarbons. Dispersive growth in water (2, 15-18 CLW) can also become a site for degradation and assimilation.

If hydrocarbon seepage passes into an oxidative atmosphere, it faces two secondary consormial barriers. The first is at the water–air interface where 1, 19-06 FOM consorms generate oxygen -rich conditions within which the hydrocarbons are accumulated, degraded, or assimilated. If the hydrocarbons enter the atmosphere as gases, droplets, or volatiles, clouds (1, 22-03 CLD) form the last natural barriers before the hydrocarbons leave the planet.

These observations suggest that oil and natural gas are renewable resources following the pattern of production by the oxidative biomass in the surface biosphere when it descends into the reductive regions of geological media saturated with water to form biosynthesized hydrocarbons. These hydrocarbons over time are likely to reemerge from the reductive zone into the oxidative zone where degradation and assimilation occur; the product carbon reenters the biospherically driven cycle of the surface.

11.14   BACTERIAL INTERCEPTION OF GROUNDWATER FLOWS IN POROUS AND FRACTURED MEDIA

Groundwater is a major hidden resource that must be extracted to be utilized. The usual technique is to set extraction wells into porous or fractured media possibly with a pack; water then flows into the well through slots or perforations. The wells present a large surface area over which the groundwater will flow faster and faster toward the bore hole. At the same time oxygen enters the groundwater from the bore hole water column, possibly moving down the casing outside the borehole or produced electrolytically by the motors. This oxygen creates one or more oxidative–reductive interfaces where biomass plugs can form and interfere with groundwater flow. The net effect is compromise of the groundwater flow by the biomass plug that eventually closes off the well:

Mummy, Mummy, our well has gone dry!
There, there, my darling does not cry.
Daddy will come and use his gun
And blow away that nasty scum.

Figure 11.13 defines the bacterial consorms known to interfere with the unrestricted flow of groundwater by generating biomass that plugs the voids and fractures, preventing flow into wells. Originally these "clogging" events were considered chemical and physical. Over the past 30 years, the bacteriological influence that plugs wells, primarily through the growth of biomass plugs has been accepted reluctantly.

The development of a plugging can best be determined by periodic (e.g., monthly) analysis of the output of a well in service. Indications of the formation of a biomass plug in a borehole include cloudiness caused by dispersive types of planktonic growth (2, 15-07 CLS) followed by suspended biocolloidal growths seen as floating masses caused by 2, 15-18 CLW and further clouding. As the biofouling (plugging) is firmly established, the water will become turbid due to the growth of 2, 15-13 TBD. When examined in a test tube, the water appears very cloudy and dense colloidal forms may be floating in the water. On rare occasions, the water can become gel-like; this increases its viscosity (higher fcP). The probable cause is the growth of 2, 21-22 SLM slime consorms in the biomass plug that shears and sloughs into the flowing groundwater.

High organic carbon content in reductive waters may trigger the growth of silt (2, 11-16 SLT) that acts as bioaccumulators for all the passing debris (e.g., clays and fine sands). Any reduced soluble ferrous iron forms will be oxidized to insoluble ferric forms.

Two major growths associated with this iron oxidation are ochres (3, 18-19 OCR) and iron pans (3. 22-18 IPN), both of which form at the oxidative–reductive interface and may become embodied with geological strata. High calcium content causes the formation of bioconcretions (3, 18-25 CCR) that resemble living forms of concrete, particularly under alkaline conditions. These consorms can integrate into a single biomass plug at the interface; 2, 19-21 PLG will be the dominant consorm if the groundwater has a high organic carbon content; 3, 19-21 PLG will be dominant if the biomass contains iron accumulations or concretions.

The biomass plugging of a borehole can also involve back pressures when groundwater follows the lines of least resistance. These diversions may involve slime conduits (3, 10-10 SBL) that can lead to surface discharges of groundwater where it is least expected (e.g., sand levees). The sand will appear to "boil" as it relieves the back pressures around the plugged bore hole.

Two other symptoms that mark water well plugging with biomass are the formation of a sheen on discharged water (2, 22-07 SHN) and the formation of foam (1, 19-06 FOM) under significant stresses. Foaming indicates a very active eutrophic biomass plug around a bore hole. Foam in a treated well may be symptomatic of a successful disruptive–dispersive treatment to restore flow.

Ice is not normally perceived as a problem in the extraction of groundwater from a well. Under unusual conditions, ice may form in and around ochrous deposits when the temperature is below 7°C. Bacteria convert liquid water into crystalline ice (1, 16-12 ICE). Interestingly, gas hydrates (5, 15-10 GHY) generate crystalline methane–ice water structures at these temperatures at 800 to 1,500 meter depths. This ice-locked natural gas (methane) is considered the largest carbon energy reserve on the planet, exceeding the known oil, coal, and conventional gas reserves.

The sizes of the tundra regions and continental shelves where gas hydrates are formed provides a clue as to the immense scale of these locked energy reserves.

Gas and oil wells can also be subjected to biomass plugging in the waters associated with the producing wells. The plugging can cause step-wise reductions in flow that may be perceived as the failure of a well to extract. In reality, the reserves may have been plugged by 3, 19-21 PLG consorms. Decreasing productivity of oil wells is attributed to insufficient water, creating an environment that is not supportive for bacterial activities. However, growing evidence indicates that bacteria (particularly 4, 05-25 BAP) can "mine" water directly from the oil and generate dense asphaltenes (black "goop").

12 Natural Bacteriological Consorms

12.1 INTRODUCTION

Nature is awash with many bacteriologically dominated consorms, particularly in water-saturated reductive regions where bacteria dominate to the exclusion of virtually all eukaryotic life. This is because the bacteria under such conditions can function anaerobically (without oxygen) or generate sufficient electricity to cause the releases of oxygen from the water by electrolysis. Clearly the geological crust of Earth is electrically active from geomagnetic interpolar radiation.

In the natural environments that surround active biomass, oxidation–reduction potential and the local water viscosity are linked and biomasses are influenced by the fundamental pathways descending and ascending from the Earth's crust. In simple terms, the descending fronts are dominated by descending organic carbon and the ascending fronts are formed by a combination of gases, volatiles, and chemicals dissolved in water.

This chapter addresses some examples of bacteriologically influenced consorms found at various natural interfaces. Each example uses laboratory and field data produced at the research facilities of Droycon Bioconcepts Inc. of Regina, Canada, to develop methods to differentiate bacterial consorms. Emphasis is on the less understood and documented consorms. Despite the lack of information, these consorms play major roles in worldwide economic activities. Thus, these examples are of prime interest to engineering and environmental sciences.

12.2 1, 22-03 CLD (CLOUDS)

Clouds are visible masses of condensed water vapor floating in the atmosphere above the ground. They are so ubiquitous that their significance is generally related to the generation of rain, snow, and shade and the production of storms than it is to their status as dynamic living consorms. In the early 20th century, photographer Alfred Stieglitz, through his gallery and *Camera Work* journal, introduced European avant-garde art to America. He began photographing clouds using a Whatman #40 red filter. His techniques revealed textures and forms in the clouds and generated considerable public attention and many exhibitions. His works showed clouds as complex, ever-changing masses of condensed water.

A century later, the nature of the bacterial activities leading to the bionucleation of water in clouds remains to be firmly established. The HAB BART™ system with VBR72 software (Droycon Concepts Inc., Regina, Canada) routinely records very active bacterial populations (estimated at 10 to 30 million predicted active dells per milliliter during a downpour of rain. In these cases the reaction pattern is a UP reaction, indicating that the bacteria are aerobic. RASI-MIDI identifications indicate that these consorms are consistent in FAME composition and an experimental library has been established. Using the Steiglitz filter technique, black and white film images of clouds that appeared to have wooly textures (Plate 12.1). The images revealed constant changes in form that implied a very dynamic function perhaps influenced by bacterial activity.

Rusticles (3, 19-26, RST) on recovered specimens taken from the *RMS Titanic* in 1996 and 1998 were examined with X-radiographic techniques and revealed similar forms (Plate 12.2). While rusticles are very solid with densities ranging from 1.4 to 1.8 and very high iron content (85% of dried weight), radiography revealed that the iron was unevenly distributed throughout the rusticle body and appeared to be clustered in forms that resembled clouds. This means that iron in the bodies of the rusticles was manipulated in the water environment, possibly similar to the manner in which water is retained in clouds floating in the atmosphere. These physical mechanisms may arise from the electrically active binding properties of the extracellular polymeric substances (EPS) released by the bacterial cells during metabolic activity.

One major difference between the clouds photographed with Steiglitz's technique and the rusticles is the presence of linear and branching ribbons (arrows) found throughout the rusticles. Dissection revealed that the ribbons had high iron content, were typically flat, and their widths were 15 to 300 microns. These distinctive

PLATE 12.1   Saskatchewan cloud photographed using Steiglitz method.

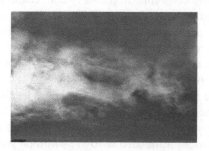

PLATE 12.2   X-radiograph of rusticle recovered from *RMS Titanic* in 1998.

components could form a structural support for electricity or some type of primordial nervous system. A rusticle may be considered an advanced form of bacterial consorm exhibiting stable, slow growth containing multiple microbial consorms active at separate sites. The clouds in the photos do not reveal the structural interconnectivity of the rusticles and clouds have much shorter life spans.

12.3   1, 16-12 ICE (ICE)

Ice has historically been treated as a solid state of water produced when liquid water freezes. Little attention has been paid to the potential of bacterial EPS to influence the physical state of water by transforming it from a liquid to a frozen state (see Plate 12.3) at temperatures above the freezing point of water. We have known since 2001 that bioconcretions (such as ochre, 3, 18-19 and rusticles, 3, 19-26) can generate ice crystals at temperatures ranging from 0 to 7°C.

The ice formed in association with concretion assumes a variety of shapes that partially result from the nature of the consormial growth. Ice may even form within a concretion, then break free into the surrounding water-dominated medium. This ice results from binding of water into the EPS in a manner that allows the water to remain frozen even at temperatures above freezing. Of particular note is the melting temperature for this EPS-formed ice. When the temperature of isolated bacteriologically synthesized ice (bio-ice) was elevated during investigation, the temperature reached 18 to 25°C before the bio-ice melted. The highest trigger temperature for melting was 55°C. These observations suggest that the boundary between frozen (solid) and liquid water is transitory in nature and can be manipulated both ways by consorms. 2, 16-12 ICE causes the transition point to move upward (EPS solidifies the water) or downward when the EPS triggers an "antifreeze" action and keeps the

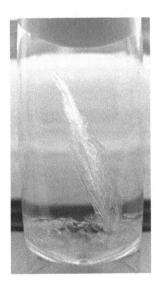

PLATE 12.3   Fully matured ice crystal generated at 7°C using a bioconcretion (3, 18-25 CCR) in 40 ml water and occupying 7% of volume.

water liquid below its normal freezing point. This phenomenon of high-temperature ice formation may partially explain why engineers and water treatment operators find ice in water pipes placed well below the frost line (2.7 meters). The pipes may have frozen from within by the activities of ochres (3, 18-19) or concretions (3, 18-25).

12.4   3, 18-25 CCR (CONCRETIONS)

Concretions are growths dominated by inorganic structures included within a growing biomass. The forms of concretions depend on the local environment and salt concentrations in the water. Plate 12.4 shows an unusual form of concretion that developed in 12% (salt) groundwater. A vertical crack in the inner vial of the BART tester allowed the CCR to creep outward into the gap between the inner and outer vials where much of the concretion then occurred. This CCR formed as a pensinsulated radiational growth with two connecting columns that crossed the gap between the inner and outer vials (arrows). Secondary growth appeared on the inside wall of the outer tester. This CCR exhibited a very distinct form and structure. CCRs usually do not form so symmetrically; they usually form randomly as do encrustations.

CCR may grow over a surface to form a patina (coating or mat). This type of growth is commonly seen on marble statuary and may gradually generate discolorations and pitting of the supporting stone or masonry. Patinas are gaining attention as major problems in fiber-reinforced concrete pipelines that carry water. Plate 12.5

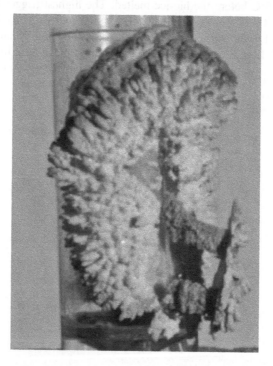

PLATE 12.4   Bioconcretion formed between inner and outer vial when IRB BART was performed on 12% saline groundwater.

PLATE 12.5 Vertical cross-sections through a fiber-reinforced concrete wall showing early impacts of consorms active in the patina.

shows a bacteriologically dominated patina that coats the surface of the concrete. The entire inside surface of the pipes appears to have been coated by the manufacturer but the coating is in reality a biomass.

Patina develops as complex structures driven by the bacteria that cause erosion of concrete but do not affect the recalcitrant fibers such as asbestos that strengthen the concrete. In general, acid-producing bacteria (APB) are active in these patinas. The APB would live deep within a patina under the reductive conditions present there and function fermentatively, releasing fatty acids as primary daughter products. Such release reduces the local pH into the mildly acidic range (3.5 to 5.5), causing the concrete to weaken and begin to spall (shear away or dissolve). The reinforcing fibers in concrete would not be affected by the erosion. They would protrude from the concrete (arrow) and become integrated into the patina forming above the collapsing surfaces.

As the patina develops, more fibers emerge from the corroding concrete. The combined effects of water flow patterns and the activities of 3, 18-25 CCR and 2, 24-13 TCT cause the fibers to become interwoven by hydraulic flow patterns and bonded by the action of EPS.

Microscopic inspection of the patina reveals a complex coating (Plate 12.6) of significantly porous EPS (arrow) but maintains a tight and durable structure. Such a "roof" over the corroding reinforced concrete surface thus provides protection to the biomass forming at the corroding concrete–water interface. The patina also limits the movement of dissolved oxygen in the flowing waters above the patina to the interface so the interface is most likely to become reductive. These conditions allow fermentation to dominate and generate fatty acids with more acidic local conditions under the patina.

12.5   3, 18-19 OCR (OCHRES)

Ochres consist of bioconcretious outcroppings of surface coatings with high ferric iron content. Unlike patinas that are relatively smooth, ochres have very irregular surfaces (Plate 12.7), in part because the bacteria tend to bioaccumulate artifacts

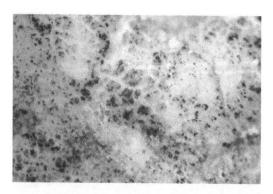

PLATE 12.6   Photomicrograph (×40 magnification) of surface of patina growing on fiber-reinforced concrete.

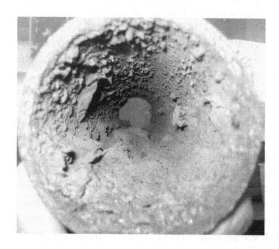

PLATE 12.7   Two-inch mild steel water pipe displaying encrustations with ochres.

from the water environment and because the form of the growth is peninsulate, extending outward from the attached surface into the water.

Eventually water pipes plug the ongoing generation of ochrous (3, 18-19 OCR) biomass. Ochre internal structures are complex, reflecting the complex biocrystallization of ferric oxides and hydroxides (Plate 12.8). These crystallized structural elements support the growing ochres and allow complex three-dimensional forms to develop even in flowing waters. Ochres form at oxidation–reduction interfaces and then develop bacterial communities. IRB, HAB, and SLYM dominate the surface coatings. APB and/or SRB communities dominate underneath. The IRB communities generate the mostly ferric iron crystalline frameworks that provide structure to ochres. Corrosion occurs via APB or SRB. The dominant community determines the nature of the corrosion.

Corrosion tends to disperse laterally, causing "dishing" in mild steel pipe walls when APB communities dominate. This is most likely when the water flow carries a high organic burden that is incorporated into the ochres while the APB communities generate fatty acid daughter products. The wall pH may drop between 3.5 and 5.5.

PLATE 12.8   Photomicrograph (×40 magnification) of sheared peninsulated ochre outgrowth from encrusted mild steel water pipe showing complex crystalline cortical structures.

SRB communities are likely to dominate under two conditions: (1) significant sulfate content (>30 ppm) in the flowing water; or (2) the ochre has a very active biomass generating significant inorganic crystalline product. Corrosion caused by SRB communities arise from the generation of hydrogen sulfide that triggers electrolytic corrosion of steel through deep lateral pitting with rapid perforation of pipe walls.

The nature of the SRB communities defines the most likely type of corrosion. High sulfate waters with low available organic loadings are most likely to generate a BB (black base) reaction in the SRB BART tester, indicating that the SRB bacteria are covert and do not form a major part of the ochrous consorm. If, however, the waters carry significant available organic loading that triggered a very active biomass in an ochre, SRB are still likely to be involved in generating hydrogen sulfide but the ochre is dominated by the degradation of sulfur amino acids in the biomass proteins.

The SRB reaction pattern would be BT because the jet black sulfide first forms around the lower half of the BART ball. A BT reaction indicates the probability that electrolytic corrosion would be generalized with more lateral pitting. When active APB and SRB (BT) are communities, lateral corrosion may be relatively fast and involve increased porosity of steel pipe. This more widespread corrosion will prevent a pipe from maintaining pressures when applied and allow it to "weep" water from the microperforations. These perforations form where the lateral corrosion extends into the steel. It is likely that the steel will be subjected to fresh growth of ochrous biomass over and around the perforation. Biomass growth over microperforations is likely because the perforations create an interface between the ochrous biomass and the environment outside the pipe. Ochres can be grown using an Ochre Mesocosm Apparatus (OMA, Droycon Bioconcepts Inc., Regina, Canada). In principle the OMA creates high surface area to void ratios at multiple oxidation–reduction interfaces in a high iron environment. The net result is ochre (Plate 12.9) as a biomass in the form of encrustations. Preventative maintenance and treatment strategies utilize OMA to determine the most suitable management strategy.

To culture ochrous biomass, the OMA includes a recycling air-saturated system enriched with culture media formulated for IRB communities. One or both chambers

PLATE 12.9 View along a 3-inch polycarbonate secondary OMA chamber showing unstructured ferric-rich ochrous biomass.

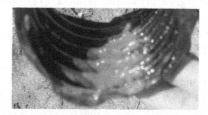

PLATE 12.10 Photograph of a 6-inch section of a slotted ribbed plastic drain containing copious ochrous growths and deposits 6 months after installation.

can be filled with the media pack of interest (e.g., sand, gravel) or the culturing fluids can be passed through spaces of potential geometric significance. The configuration of the passageway as a narrow rectangular slot or as a circular hole can be compared. The void-to-volume ratios in the pack can be compared to determine the most suitable pack material that would inhibit ochrous biomass generation.

Ochres are particularly problematic in ribbed flexible plastic drains carrying geotextile stocks, particularly in regions where the groundwaters are high in iron. Ochres can grow on the geotextile materials, in the slots, and in drains (Plate 12.10).

Weeping tile drains are constructed of ribbed polyethylene black ribbed plastic. The material is slotted to relieve water pressures created during high flow conditions around a house, relief dam, or other structure by diverting water flows into the drain. Unfortunately these barriers do not prevent ochrous growths that cause drainage system failures due to plugging of the slots and biologically influenced impermeabilities in geotextiles. These events lead to failure of a drainage system to relieve water pressure by draining off surplus water. The result may be flooding of buildings and/or collapses of other structures such as earth-filled dams and levees.

PLATE 12.11 Ochrous biomass forming at base of dam near piezometer.

Ochres are natural formations generated where waters with high iron content pass through an oxidation–reduction interface. Plate 12.11 shows an ochrous growth around an outflow of groundwater seepage from a rock formation and also from a piezometer. This growth is a bright orange-brown outcropping emerging from the piezometer pipe, supplemented by local seepages. Over time, the ochres will mature to darker brown and ferric iron will reach 70 to 85% of the dry weight of the biomass. Surface outpourings of ochres may represent: (1) a local event—the groundwater feeding the growth is reductive; or (2) deeper ochrous plugging in the groundwater. In the latter case, the environment of the groundwater must also be shifting to oxidative.

Ochre has high ferric iron content in its biomass. The iron increases with growth, reflecting uncontrolled bioaccumulation. Ochres are symptomatic of plugging and corrosion problems. Two approaches may be taken in developing strategies to manage ochre risks: (1) preventative design and regular maintenance; and (2) regenerative treatment when a problem is significant and not easily treated. Ochre treatment is challenging because ochre is very robust and effective treatment must be strategic. No chemical is likely to address all the issues or a chemical may produce consequences too extreme for routine use. Additionally, natural ambient temperatures are generally not conducive to effective treatment and the application of heat as a part of ochre management has not been effectively introduced as a routine procedure. Of the various methods for culturing ochres as a step to managing them, A glass bead pack medium allows the ochres to be easily seen (Plate 12.12). When the growth is matured, the clear plastic column appears dark and the contained voids exhibit at least 30% ochre content.

Under acceptable cultural conditions ochre growing within a glass bead vertical column can be treated by a combination of chemicals and heat. One treatment strategy involves a chemical treatment combining a biocidal detergent (CB-4, ARCC, Port Orange, Florida) and glycolic acid. When these chemicals are applied together, the ochre is shocked, disrupted, and dispersed within 2 hours, leaving the glass beads clearly visible (Plate 12.13). Such treatment provides clear evidence of the potential effectiveness of this type of ochrous condition control.

PLATE 12.12 Two-inch polycarbonate tube filled with half-inch clear glass beads and voids filled with growth of ochrous material.

PLATE 12.13 Post-blended chemical treatment applied to ochrous column (see Plate 11.12) after 2 hours of circulating chemicals.

12.6 3, 06-24 PTG (PITTING)

Pitting of solid surfaces (raw, coated, or cathodically) of mild steels is of considerable concern because pitting is a precursor of perforation and structural failures. Pitting begins when the 3, 18-19 OCR consorm forms encrustation (Plate 12.14) high in ferric iron at the oxidation–reduction interface, generating highly porous crystalline structures that have irregular surfaces that increase contact with the flowing water by channelling and increasing the surface area. This causes greater contact with the encrustation.

3, 06-24 PTG formation commonly occurs in the lower third of a vertical profile encrustation where the oxidation-reduction interface becomes reductive and stimulates the two bacterial groups associated with acidulolytic or electrolytic corrosion. Acidulolytic corrosion arises from active fermentation in the biomass that produces sufficient fatty acid daughter products to lower the pH and reduce the structural integrity of the underlying steel. Electrolytic corrosion results from the reductive production of hydrogen sulfide that instigates the process. Both PTG activities cause erosion of a steel surface (Plate 12.15) as a series of lateral dishes in the steel.

Lateral dishing is unevenly spread and reflects the amount of PTG activity in the overlying encrustation on the reductive side of the vertical profile across the biomass. It is possible to emulate the PTG consorm by modifying an SRB or APB BART to

PLATE 12.14   View of an 8-inch encrustation growing in slow-flow mild steel water line.

PLATE 12.15   Half section of an 8-inch mild steel water pipeline suffering from lateral corrosion causing dishing.

allow direct contact between the culturing biomass in the tester and the steel surface. A ¾ inch rubber washer from a garden hose was sterilized and placed on a steel coupon (Plate 12.16). A ¼ inch hole was drilled through the cap of the tester containing the culture while the tester was held upside down (see Plate 12.24). The washer was placed over the tester base and the coupon bolted onto the tester before uprighting the tester. If adequate pressure was applied, the culturing biomass in the tester would remain perched over the steel and corrosion could be monitored. This experiment took 3 months under enhanced laboratory-based conditions.

Pitting is not necessarily a surface corrosion phenomenon (lateral dishing). It can also intrude into steel and form cavities in steel pipe walls (Plate 12.17). The section illustrated was found in a stainless steel 8 mm thick; the cavity penetrated 60% through the wall (4,800 microns) and extended 6,720 microns laterally. The gateway (portal) into the cavity was only 28 microns away at the closest point (white arrow). This confined biomass exhibited considerable differentiation of the types of structures observed. Of particular

PLATE 12.16 Pitting generated in mild steel held beneath an APB BART device culturing an APB-dominated biomass.

PLATE 12.17 Photomosaic section showing biological structures inside corrosion cavity in stainless steel showing narrow portal of entry.

PLATE 12.18 Cross-section of corroded concrete resulting from activities of APB communities within a PTG consorm.

note was the "roof" of the cavity (black arrow) that appeared thin and likely to collapse. Such a collapse would expose the biomass to the outside environment and convert the cavity into a deep pit that would lead to perforation of the stainless steel wall.

Cavity formation as a corrosive event can also occur in concrete (Plate 12.18). The vertical section came from concrete that had been supporting a patina (removed). A number of deep peninsulated cavities (pits) were driven 2 to 8 mm into the concrete. They appeared to have been generated by APB as a part of the 3, 06-24 PTG consorm. The APB subjected the concrete to limited mild fatty acid impacts localized where the APB communities were most active.

12.7 3, 03-19 PFR (PERFORATION)

Perforation is the result of piercing a material and creating a hole. It can involve the dissolution or disintegration of the material by bacterial consorm activities. The usual sequence for perforation is (1) surface growth, (2) penetration, and (3) finally perforation. One consormial combination capable of this 3, 18-19 OCR followed by 3, 06-24 PTG, then 3, 03-19 PFR. The initial ochrous formation leads to pitting and subsequent perforation. Plate 12.19 illustrates the process. A relatively smooth growth of encrustation appeared on a stainless steel surface; then perforation occurred (arrow). The hole through the steel and the encrustation became coated with a single, thickly viscid 1, 19-06 FOM bubble marking the spot where perforation occurred. The bubble effectively blocked the hole, preventing leakages through the hole until the foam bubble plug disintegrated. This phenomenon is another of Nature's Band-Aids; it prevents leakages (pressure losses) until the biomass forming the viscid bubble collapses.

Certain perforations of steel are not associated with massive ochrous growths and pitting. Plate 12.20 illustrates a perforation through the steel floor of a gasoline storage tank. Surface pitting by 3, 06-24 PTG (white circle) was visible but no evidence of the active biomass was found when the failed plate was removed.

Microbiologically influenced corrosive processes terminating in perforations involve a sequence of growth and maturation activities of 3, 18-25 CCR, 3,

PLATE 12.19 Encrustation growing over stainless steel surface in which pitting coated with a viscid foam bubble appeared.

PLATE 12.20 Perforation in steel floor of gasoline storage tank.

06-24 PTG, and 3, 03-19 PFR consorms. Essentially the biomass grows and becomes more active in the first two stages (CCR and PTG). The greater amounts of activities allow the detection of adenosine triphosphate (ATP). Drainage through a perforation can trigger a radical reduction in the biomass. This means that ATP levels during the formation of a perforation reflect a high level of biomass activity in the CCR and PTG phases. Several perforations may appear with a pitting of biomass clusters (Plate 12.21). ATP levels would be high as these coupled biomasses create perforations. The plate shows one perforation of the mild steel casing with three possible sites where radical pitting is present and perforation could occur (arrows).

Such perforations resulting from multiple adjacent pitting events occur when a broadly based biomass generates a series of dishing activities in close proximity. Perforations can also be very localized when steep corroded walls form around them (Plate 12.22). Such isolated perforations away from other evidence of pitting indicate electrolytic rather than acidulolytic corrosion produced by a focused active biomass. The dominant bacteria in the biomass can reductively generate hydrogen sulfide from sulfate (BB reaction) or sulfur amino acids (BT reaction).

PLATE 12.21   Single perforation of mild steel wall with three pitting zones that may become perforations.

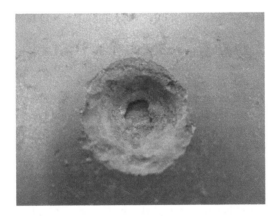

PLATE 12.22   Vertical steep sloped 3/8 inch perforation through mild steel plate showing steep walls indicative of bioelectrolytic corrosion by hydrogen sulfide.

These conical perforations are more frequent where a focused biomass develops locally. Anodic charges impressed in the steel may also generate a perforation event. These charges can be attributed to geological pack material of high iron content (.e.g., magnetite, goethite).

Severe physical stresses to steel pipes can set up embrittlements that allow pitting consorms easier access to the interiors and cause lateral perforations in the form of lateral splits in the pipe walls (Plate 12.23). In this case, the splitting of the steel was accompanied by splitting of the concretion along the fracture line.

12.8   2, 09-15 MIC (MICROBIOLOGICALLY INFLUENCED CORROSION)

The bacterial activities that cause microbiologically influenced corrosion are defined separately in the atlas. This type of corrosion requires a dispersed biocolloidal or

PLATE 12.23   Fractured lateral perforation of steel pipe coated with concretious growth that failed at the same time the steel wall cracked.

relatively structured biomass. Normally, this biomass would be very metabolically active and the easiest way to start assessing corrosion risk is determining ATP levels in the biomass. Sampling may be challenging if the biomass is dispersed (biocolloidal), in which case a water-dominated sample would be required. A sample can be removed from a defined biomass on the surface of interest.

Keep in mind that a biomass structure is likely to involve oxidation–reduction gradients and the corrosive activity would likely appear on the reduction side of the gradient. Sampling within a biomass is preferable to simply taking a surface sample, while ATP tests such as those produced by Luminultra Technologies (Fredericton, New Brunswick, Canada) can indicate levels of metabolic activity in a sample. ATP activity is a significant factor in determining risk from microbiological activity but ATP testing does not indicate the type of corrosion that may be expected.

Three BART methods can accomplish this by measuring the activities of three critical groups causing this type of corrosion: SRB, APB, and HAB testers. SRB BART reactions determine the levels of risk for electrolytic corrosion from hydrogen sulfide. The BT reaction reveals that a larger biomass may be involved, with hydrogen sulfide generated reductively from the sulfur amino acids in the biomass. If the hydrogen sulfide is produced through the biological reduction of sulphates, a BB reaction will be indicate a greater risk for radical perforations of steels. APB BART analyses detect reductive fermentative activities common in larger biomasses attached to steel and concrete surfaces. Interfacial pH will drop into the 3.5 to 5.5 range of 3.5 to 5.5—enough to stimulate acidulolytic corrosive processes.

HAB BART detects bacterial activities that break down organics (important at many bioremediation sites). The generation of UP reactions indicates small potential for corrosion but high potential for biomass growth. DO reactions indicate a bacterial biomass functioning anaerobically under reductive conditions. The potential for corrosion of steel is more extreme with 3, 18-25 CCR; 3, 06-24 PTG; and 3, 03-19 PFR consormial sequences. For concretes and grouting materials, DO reaction from HAB coupled with a DY reaction from APB BART indicates potential for corrosion.

Examining the activities of MIC-related events in a laboratory setting is a challenge. Two critical activities are: (1) accelerating the corrosive processes; and (2) developing a strategy to restrict or prevent corrosion events from becoming serious management issues. One simple adjustment of BART apparatus is inverting the tester and drilling a hole through the cap to create contact with the surface of concern. Plate 12.24 demonstrates the potential accelerate the rate of PTG to PFR corrosion in s steel coupon already suffering from minor pitting. The inverted tester provides a confined selective nutrient medium to be inoculated by the indigenous bacteria of corroding steel surfaces. The extremely reductive conditions in the inverted cap confined by a pressure-sealed washer accelerate the rate of corrosion via PTG and PFR consormial events.

MIC constitutes microbiologically influenced destruction of solid surfaces. The biomass generated often takes the form of a ferric iron-rich concretion (Plate 12.25). Concretious growths form obvious coatings. They also harbor the potential for corrosion within the reductive regions and affect equipment functioning by increasing mechanical wear and reducing efficient water movement through the impacted sites.

PLATE 12.24   Close-up of BART corrosion-testing apparatus.

PLATE 12.25   Badly coated pitless adapter no longer functions effectively because of concretious growths.

PLATE 12.26   Golf green suffering dieback and balding as a result of black plug layer competing successfully with turf grass in Regina, Canada.

12.9   3, 10-21 BPL (BLACK PLUG LAYERS)

Black plug layer biofouling develops in porous media such as sands, sand–clay mixtures, and lighter soils that have high porosity and permeability and are used for crop production. One such "crop" is turf grass used on golf course greens. Management of these greens involves a combination of high levels of fertilizer application to stimulate grass growth and mowing the grass very short. The net result is that while the turf grass biomass is stimulated by the nutrient applications, constant mowing to the desired length creates stress. Microbial biomasses in soil- or sand-based greens are also stimulated by fertilizers, particularly when applied to the greens and not to the foliage. A microbial biomass can better compete with grass roots when the plants are under stress from the constant mowing. As a result, the microbial biomass becomes dominant over the turf grass and causes the grass to die (Plate 12.26).

The 3, 10-21 BPL consorm forms an impermeable layer just under the surface of an infested green. The layer prevents access to water by the deeper roots by lateral plugging of the sand or soil just below the surface. The initial event is a chlorotic effect (yellowing) in the grass followed by dieback when the grass can no longer compete. Plate 12.27 shows the central swath of a golf green denuded by BPL infestation. Attempts of the greens keepers to plant fresh grass seedlings failed. Impermeability in the infested part of the green was easily demonstrated by throwing four gallons of water over the area. The water would not penetrate the green, ran off to the sides, and drained into the ground only when it reached healthy grass.

Coring the impacted green showed the black plug layer just under the surface. Plate 12.28 shows a core from a green containing BPL consorms that were very active but had not yet totally dominated the grass. A layer of black growth can be seen at the top of the core (arrows) with deeper set dark brown or black structures deeper in the core.

Cores removed from BPL-infested golf greens show two signs of risk: (1) formation of a jet black layer just under the surface; and (2) lack of grass roots in the core; the core maintains structure due to the thick slimes that bind the media together (Plate 12.29). BPL infestations are common in sand greens, particularly when fertilization is applied directly. This practice allows the BPL to "feed" on the fertilizer

PLATE 12.27 Green severely infested with BPL that caused denudation of central swath; attempts to plant gras seedlings (arrows) failed.

PLATE 12.28 Core sample from green suffering from black plug layer infestation that did not cause dieback or severe stress.

PLATE 12.29 BPL-infested core sample showing lack of grass roots while retaining structural integrity due to high slime content in sand core generated by BPL.

directly and generate biomass. Foliar application of fertilizer presents the advantage that the plants can take nutrients directly through the leaves.

Remediation of a 3, 10-31 BPL-infested green can be achieved by changing the conditions that support the infesting bacteria in the layer. Two key factors in infestation are: (1) very reductive conditions become due to the plugging that kills the grass roots; and (2) application of a solid fertilizer directly to the green stimulates the activities of BPL consorms. Another corrective step is aerification of the green to break up reductive zones and allow the roots to penetrate the soil more easily. Hydrojetting aerification injects oxygenated water into a green to enhance the oxidative state. Another effective corrective technique is applying fertilizer as a liquid aerosol so that it directly contacts the grass leaves and bypasses the sand or soil.

12.10   3, 10-27 BBR ("BLUEBERRIES")

Little has been published about the spheres that are generated during the growth of concretious biomasses. These spheres are particularly common in 4, 15-25 BAP and 3, 19-26 RST. Microscopic examination commonly reveals clusters of spherical objects (Plate 12.30) that exhibit a wax-like appearance. Isolated spheres (Plate 12.31) have high iron contents. Clustered spheres may serve as energy reserves for the biomass. The functions of dense, isolated spheres with high iron content are more debatable and may reflect a mechanism to isolate surplus iron in the biomass or relate to electrical potential within the biomass.

These iron-rich spheres common in 3, 19-26 RST and can create problems when a growing biomass attaches within a pipeline environment. The hydraulic flows in the pipeline loosen the iron-rich balls that then move through the pipe, channelling and gouging the steel floor. Over time, the result is thinning of the gouged floor and the escape of liquid from the pipeline.

These spherical growths range in size from 5 to 30 microns. The origin of the "blueberry" term is the appearance on the surface of Mars of balls that appear similar to products of bioconcretious growths. The balls may have evolved when Mars had healthy water cycles and rusticle growth.

PLATE 12.30 Photomicrograph (×40 magnification) of cluster of "blueberries" in freshly recovered rusticle.

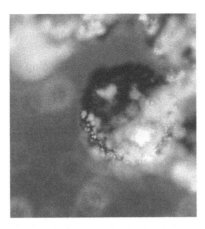

PLATE 12.31   Photomicrograph (×40 magnification) of a high iron content "blueberry" in freshly recovered rusticle from a high temperature saline pipeline.

12.11   3, 19-26 RST (RUSTICLES)

The rusticle was so named because it represents a combination of rust and icicle growth commonly found on steel ships sunken at depths of 1 kilometer or more. Plate 12.32 shows a recently recovered rusticle from a ship wreck 4 kilometers deep in the North Atlantic. Rusticles incorporate a number of bacterial and fungal communities that function separately within a common concretious consorm. The separation allows connection through water channels between the crystalline structures comprising most of the biomass. The ferric oxide and hydroxide crystalline matrix of the biomass has a high porosity (>70%) and high iron content (60 to 85% of rusticle dry weight). The iron may be replaced by aluminum or other metals if the local environment favors assimilation and crystallization of structures made from these alternative metals. Examples are steel shipwrecks incorporating structural aluminum and ships that carried ore (e.g., bauxite that has high aluminum content).

Typical bacterial communities recovered from freshly recovered rusticles are IRB, SRB, HAB, SLYM, and DN. Each occupied a different area of a rusticle. More communities appear in water channels associated with the ducts linking rusticles

PLATE 12.32   Freshly recovered rusticle 11 inches long with right side uppermost in its natural position on ship exterior.

to the outside environment. Outer walls of rusticles are generally layered and contain significant populations of fungi. Strands of mycelia (cell threads) may be seen emerging from rusticles.

Rusticles can provide archival information about ship sinkings. For example, the *RMS Titanic* sank in 1912 after colliding with an iceberg. The collision induced physical stresses in the steel. As a result, the steel hull broke open at three major points, releasing coal and debris as a cloud that followed the ship down to its final resting place. In rusticles growing on horizontal surfaces, fragments of coal and glass are concentrated inside the growing structures.

Stresses on steel hull plates cause embrittlement that leads the steel surfaces to expand and fracture, creating ideal conditions for early colonization by rusticles that tend to concentrate at zones of embrittlement. They grow only on one side of the steel. This observation then led to consideration of the influence of electrical charges on steel surfaces. Anodic impression tends to support rusticle growth and cathodic activity deters the potential for attachment and growth.

Culture of rusticles was a challenge until 1999 when the effects of electrical charges on growth became apparent through laboratory investigations. Plate 12.33 demonstrated the potential for fast rusticle growth on mild steel impressed with anodic charges. Descending growth rates up to 5 mm per hour were achieved, completely generating a ferric-rich rusticle growth in 4 days. A U-shaped mild steel plate was set vertically to trigger growth from a dispersed homogenate of rusticles suspended in sterile 4% seawater as a 10% by weight inoculum.

While the rusticles generated artificially grew at accelerated rates, they developed into regular forms incorporating the typical brown coloration of ferric forms within a biomass. This raises the interesting question of whether dissimilar metals present at a ship wreck site create electrical charges that foster rusticle growth.

PLATE 12.33 Rusticles forming and growing down from a 12-inch U-shaped mild steel structure that had been anodically impressed.

PLATE 12.34 Photomicrograph (×100 magnification) showing lateral stacking of ferric iron-rich crystalline structures that support rusticle biomass.

PLATE 12.35 Photomicrograph (×100 magnification) showing unusual pink ferric iron-rich crystalline structure revealing rings of radial crystalline structures at specific locations.

Rusticles in nature continue to accumulate ferric iron and other metallic cations until the biomass can no longer function. The site of *RMS Titanic* revealed a 7-year rusticle cycle: formation, growth, then collapse as a spent biomass, with ferric iron exceeding 85% of dried weight. Internal crystalline structures provide structural support (Plate 12.34) for a very porous biomass containing conduits and reservoirs within which the water can move and pool. Thread-like forms appeared to interconnect the various structures and crystalline forms (Plate 12.35) found mainly in water channels inside the rusticles. Microscopic investigations of many rusticles recovered from steel shipwrecks indicate that the primary crystalline structural elements give rusticles the form and ability to function as a consormial biomass.

12.12 1, 19-06 FOM (FOAM)

Foam is a matrix of bubbles formed by the entrapment of air or gas in a turbulent water-based environment. Foam is formed by waters rich in polymeric materials such as EPS and proteins that form where water interfaces with air. Foam has a natural ability to hold gases that rise through water as bubbles or enveloping gases such as air entrapped in wave actions. Foam generates where bacteriologically active water and entrapped gas provide greater surface area for the activities of 1, 19-06 FOM consorms. One frequent effect of the typical shock–disrupt–disperse scenario of biomass treatment is the loading of the treated waters with dispersed EPS, causing foaming. Plate 12.36 shows considerable foaming generated during a treatment. While the foam can be considered a detrimental side effect of treatment, it also indicates the disruption of the biomass. The foaming by-product in oil wells is very detrimental. It blocks transmission and causes wells to become unproductive for months until the foam plugging collapses.

In laboratory trials using mesocosms, aeration systems commonly generate foam capping (Plate 12.37) resulting from rampant growth of nuisance bacteria in the simulated plugging. The foam indicates that the bacteria in the mesocosm are extremely active, producing large amounts of EPS and proteins. Foam on the downwind sides of lakes indicates high levels of EPS and possibly proteinaceous products of pollution or very active eutrophic communities in the waters.

PLATE 12.36 Water well head indicating destruction of obstructive biomass and subsequent formation of foam.

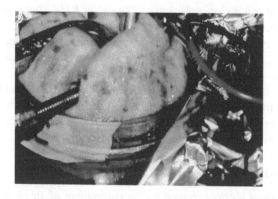

PLATE 12.37 Foam formation in treatment.

12.13   3, 15-17 TCL (TUBERCLES)

Bacterial growths may form dome-like structures to harbor the biomass. Often the interface between the dome and the air or water generates an active oxidation–reduction interface that becomes the focus for the formation of a ferric iron-rich, nodular cap (Plate 12.38). During maturation, these nodules can modify form and function and simple symmetrical domes may become more eruptive, elongated tubercles. Biomass may erupt slowly from the dome. Tubercles, like nodules, form specific corrosion sites between the biomass and the underpinning steel. Electrolytic (SRB-initiated) or acidulolytic (APB initiated) metal corrosion can occur.

12.14   3, 18-14 LSL (LATERAL SLIME LAYER)

Lateral formations of slime (Plate 12.39) grow as a series of horizontal layers that reflect sequences of events. The plate shows 3, 18-14 LSL in the base of a pipe at an industrial landfill drain subjected to seasonal variations in precipitation and chemical loading. Heavy precipitation triggered flow through the drain followed by quiet periods without precipitation of flow. The effect on the formation of biofilms was dramatic. The flush (high flow) and drain (no flow) cycle altered bacterial activities within the drain. High flow added oxygen and particles to the system. As flow slowed, the particles carried in the flow settled out and become bioconcreted into the growths on the floor.

As the surplus waters receded over the growing concretion. the next phase was the bioconcretion of the inorganic particles settling out in the drain. A highly oxidative headspace over the stagnant water effectively narrowed the oxidation–reduction interface. Much of the bacteriological activity focused on generating lateral biofilms rich in concreted particulate material. The cross-section reveals significant changes in the bioconcretions formed over time. Ascending vertically, the sequence in 3, 18-14 LSL is dense material formation, cavitations, denser material, highly crystalline

PLATE 12.38   Mild steel well casing displaying rings of nodular outgrowths.

PLATE 12.39   Cross-section (1.5 inches) through concreted lateral slime layer in relief drain of industrial landfill.

cavitations with lateral voids, dense material, thin lateral voids, and finally a layer of denser material at the top.

Plate 12.39 shows a relief drain from an industrial landfill. The discharges reflected precipitation and the contents of the water discharging through the drain. The nature of the cavitations and the densities of deposited and concerted materials reveal the nature of these contents. Cavitations probably resulted from releases of degradable organics that stimulated the growth of biomass; most bioconcretious activities occurred below the biomass. Cavities could be created by the initial formation of biomass that then moved (generally upward with the oxidation–reduction interface), leaving a void supported by the crystallized concretious elements formed above and below the site. Gases such as methane (deeper) and carbon dioxide (shallower) may have played a role in the formation of a petrified cavity.

Bioconcretious structures in lateral slime layers are formed by ferric iron forms (oxides, hydrates, and also carbonates). The forms and densities of these materials are based on the chemistries of the waters draining from land fills over time. Chemical analysis may reveal the changing natures of the fluids passing through the drain shown in the plate. Bacteriological analysis is more challenging because most of the biomass focused at the oxidation–reduction interface is mobile.

12.15   5. 15-10 GHY (GAS HYDRATES)

A significant and little-explored bacteriological phenomenon is the formation of gas hydrates (clathrates) that litter continental shelves around the globe. These growths are thought to contain twice the energy reserves of coal, gas, and oil combined. Their main content is methane, locked with water into a crystallized structure (molecular ratio 8:1 methane:water). These clathrates appear as crystallized ice forms at temperatures in the 0 to 7°C range. Essentially, clathrates contain bacteriologically generated surfactants (EPS) that are able in deep oceanic environments (1 to 3 kilometers below the surface) to cause critical crystallization. The functions of the 5, 15-19 GHY communities remain speculative. Most research relates to chemical molecular and mathematical modelling. Evidence to date indicates that clathrates function across reduction gradients in ice structures above the normal freezing point for water under the specific physical conditions, indicating that bacteria in the consorm are reductive. However, strong geo-magnetic or biologically originated electrical forces may electrolyze some of the saturated liquid and solid water associated with clathrates, generating oxygen and hydrogen that may allow aerobic activities in an otherwise reductive (fermentative) environment.

Observations of the surface of Mars near the northern and southern poles reveal structures that are very biological in form. These structures become evident when the sun is low on the horizon and creates long shadows. Unusual surface structures appeared at latitude −82.02° and longitude 284.38° in a region near the south polar ice field in the eastern region of Utopia south of Protolinus. This came with the observation of a series of circular (A and B measured) and rectangular (C measured) structures that radiated from a central axis or axially along a line. Two structures (A and B) were circular and were measured at a declared resolution of 6.79 meters per pixel. Structure A appeared roughly circular, with a diameter ranging from 1,100 to 1,200 meters. Structure B was also circular, with a diameters from 590 to 680 meters. Structure C was distinctly rectangular, 490 meters long by 290 meters wide. The sun was low on the horizon when the images were taken, creating an emission angle of only 0.26° which created long shadows over the relatively flat terrain. These structures and other circular and rectangular forms on Mars revealed branching growths from a central region, clearly resembling biologically generated, whorled, branched growth. These growths on Mars presents many challenges such as determining the biological components of structures that look like limbs extending from branches that form a central trunk. Surface conditions at these latitudes on Mars are too extreme to allow normal growth but the structures observed may be recalcitrant artifacts from a geological period when the surface biosphere was active or clathrate types not observed on Earth.

12.16 CULTURING BACTERIAL CONSORMS

Culture may in this context be considered the act of attempting to nurture a specific consorm within an environment that emulates the natural environmental conditions from which the sample containing the consorm was taken. Plate 12.40 shows a typical stacked platform for culturing 3, 19-21 PLG consorms. Each mesocosm is stacked to allow flow from the base cosm to the pressure head vessel. Aerated water and nutrients were placed into the head vessel and discharged to the base cosm. This created an oxidation regime in the base cosms, the fluids moved upward, and the oxygen was stripped off bacteriologically, thus establishing an oxidation (bottom)–reduction (top) gradient.

As the PLG cosm dominated by IRB grew in the voids of the four stacked cosms, the growths became visible through the clear plastic walls (Plate 12.41). When the voids in the gravel pack became colonized with PLG, it became impossible to see the gravel except at direct contact points (white arrow). Plugging biomass can be seen as darker brown regions (black arrow).

These types of PLG-infested cosms lose hydraulic transmissivity by one or two orders of magnitude. When significant plugging of the voids (at least 40% loss of

PLATE 12.40 Four mesocosms in stacked culture platform for growth of ferric accumulating PLG consorms.

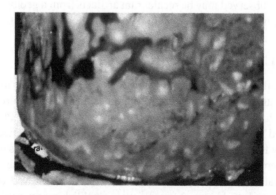

PLATE 12.41 Close-up of base cosm section showing dark ferric iron plugs formed in voids by iron-related bacteria.

transmissivity) occurs, the recommendation is to take the cosm off-line and conduct treatment validation studies to define the best protocol to effectively remove the biomass and return the flows to original values.

Cosms can also be set up to examine other plugging phenomena. Plate 12.42 shows a coupled four-cosm system utilizing sand as the porous medium. Each cosm

PLATE 12.42 Four coupled in-line cosms demonstrating potential to prevent natural gas flows by induced bacteriological plugging in sand-based porous medium.

has a different concentration of seawater salt (0, 4, 8, and 16%). All four cosms were then coupled into a natural gas line and arranged so that each cosm could be placed in line to flare gas by using a Bunsen burner.

The objective was to demonstrate whether significant plugging could occur in natural gas flowing through porous media. To accelerate the potential rate of plugging an equivalent of 10% of the void volume of each cosm was inoculated with an equal mixture of IRB (BC reaction), HAB DO reaction), and SRB (BT reaction). Natural gas was charged through the cosm and discharges were limited to a 2-hour flare for each 24-hour period. Environmental conditions were reductive; the gas phase was dominated by methane from natural gas. Discharges were monitored by the height of the Bunsen flame and also from a standing head discharge rate. Within 17 days, the 8% salt cosm was 60% plugged and the 4 and 16% salt cosms showed 35% declines in flows using the standing head method. The cosm lacking salt underwent minor plugging; it performed on day 17 as well as it performed on day 1. This indicates that the porous media through which natural gases flow can be plugged with 3, 19-21 PLG even in extreme reductive environments. Crude Bunsen burner flame heights also correlated with losses in transmissivity of water through the cosm. These initial findings indicate the vulnerability of gas wells to bacteriological plugging even under conditions that would previously be considered too extreme to support microbiological activities.

12.17 REHABILITATION

The generation of biomass and daughter products such as gases and slimes can impact the transmissivity of groundwater through porous media and fractured rocks. When this growth becomes significant, it causes resistance in flows due to the growth of the biomass and diversion of the water paths from the spring or well. Generally, the net effect over time on production from (extraction) or into (injection) a well occurs in a step-wise manner. Eventually the performance of the well or spring will be severely impacted to the point of becoming a dry hole (no flow). The time required to develop such biomass plugging depends on the local environment. Because an extraction or injection well is a capitalized investment and the service (producing or injecting water) has value, plugging seriously affects the economic management of a well.

Two standard approaches to management of biofouling wells are based on the ability of biomass to plug (occlude) a groundwater system, preventing flow or impacting water quality—making the product unacceptable. The objective of management is to render a well sustainable so that it continues to function economically over the long term. Costs are factors in the two standard approaches. A third option (ignore the plugging and replace the well when performance is unacceptable) involves steep capital costs.

The standard approaches work in harmony to maximize sustainable performance. The first is a program of preventative maintenance (PM) that reacts to any decline in well performance with a low-level treatment designed to return the well to acceptable production by reducing the impact of the biomass. While PM may work effectively for a time, the biomass causing the biofouling may rebound and cause severe failures in flow or losses in water quality. This requires a more vigorous rehabilitative/regenerative treatment (RRT) to correct the effects of biofouling.

PM usually takes the form of a simple blend of chemicals that: (1) decrease the activity of the bacteria in the biomass; and (2) disrupt some of the biomass to allow the water pathways to reopen. The chemical blend should include a biocide to kill at least some of the biomass bacteria and a detergent to break up the biomass. PM should allow a well to recover lost production and assure adequate water quality. PM of a normal extraction well involves annual treatment. If unchecked degeneration occurs, treatments may have to be delivered monthly, with time intervals extended after the PM is successful. PM is effective only when well performance exceeds 80% of its original developed capacity. PM is usually applied when decline reaches 5 to 15%.

RRT becomes essential when PM fails to regenerate performance and production drops below 80% of the original. Treatments are more complex and staged because the biofouling has become more complex and may resist PM treatment. RRT requires three steps: shock, disrupt and disperse. Shock is the application of biocidal conditions to traumatize the bacteria in a biomass. The disruptive second stage triggers the break up of the biomass as an integrated entity. It usually involves upward or downward manipulation of pH, preferably 4.0 units from the normal values, along with use of a dispersant to accelerate break-up. Dispersion removes the biomass from the site of infestation through a combination of physical force and detergent. Each biofouled site exhibits different characteristics and care must be taken to ensure the applicability of RRT.

One additional factor in RRT treatment is the application of heat. The food and chemical industries established the ability of heat to: (1) increase the rates of most chemical reactions; and (2) traumatize and kill microorganisms. Ideally the temperature rise at least 40°C to achieve both effects. A blended chemical heat treatment (BCHT™) produced by ARCC Inc., Port Orange, Florida is a heat-stimulated RRT program routinely used to control biofouling. An additional benefit is the generation of a thermal gradient around the treated well. This gradient causes differential impacts on the biofouling, causing trauma and disruption of the biomass. This improves the effectiveness of the treatment but causes prolonged "bleeding" of traumatized bacteria from the well after treatment.

Blended treatment strategies incorporating heat are effective for treating other forms of biofouling. Plate 12.43 illustrates the use of a biocidal detergent (CB-4) along with heat (increase of 45°C) on black asphaltene-rich plugs (4, 05-25 BAP) in crude oil pipelines. The pipe shown was 90% plugged with BAP. Treatment with 5% CB-4 and heat disrupted and dispersed the plugging biomass in 2 hours.

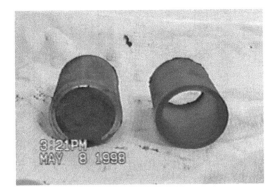

PLATE 12.43   Two sections of 3-inch mild steel pipe with left sample plugged with BAP and right showing impact of CB-4 treatment with heat, causing dispersion of black asphaltenes.

The post-treatment flows that indicate successful RRT treatment take several days to fully materialize. In extraction water wells, post-treatment discharges (Plate 12.44) should visibly demonstrate the success of treatment. If the discharge is water, early discharges commonly include dispersed biomass and the water may be clouded, colored and viscid. After the dispersed material has been dislodged and removed, the water will become acceptably clear. It can take up to 8 weeks after treatment for all the dispersed biomass to be dislodged.

Video camera inspections of biofouled regions, both under and above water, helps define the approach to treatment. Plate 12.45 is digitized image of a 4-inch horizontal well that suffered from extreme stress due to eutrophication from high nutrient loading. A 50-meter examination of the well revealed different types of biomasses. The complex interwoven biomass shown in the plate appeared to be radiating away from central columns, forming complex networks of growth. These growths interfered with the transmission of water through the well. RRT involved a combination of lowered pH and application of 0.5% CB-4. This combination disrupted the biomass that was then dispersed by physical surging.

PLATE 12.44   During early pumping, post-RRT treatment discharges can contain poor quality water and become clearer after all dispersed biomass is pumped out.

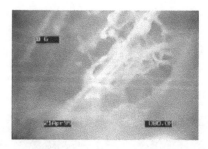

PLATE 12.45 Digital image of growths in a 4-inch horizontal well in which main thread (arrow) connects to bottom and top walls.

PLATE 12.46 Biological threads (×40 magnification) released from BAP.

Biomass in porous and fractured media and in slots and perforations in wells is a problem not limited to water wells. Evidence indicates that bacteria can also grow in oil and gas wells, causing losses in production. Traditionally such environments were considered too extreme to allow microbiological activities. However, the production losses often occur as step function declines, suggesting that the production losses are biologically instigated as plugging. In gas wells affected by decreasing production, large bacterial populations may clearly indicate a biomass downhole in the producing zones around the well. In a dry gas well, plugging occurs at some distance from the bore hole.

Oil wells are generally considered to present too extreme an environment for growth and survival of microorganisms. However, the 4, 05-25 BAP consorm is active around crude oil. Evidence indicates that BAP "mine" water from the crude oil and forming concretious coatings fabricated from asphaltenes (black "goop"). Treatment with 5% CB-4 and heating 58°C above normal causes the black goop to collapse and release debris from the BAP consorm. The debris floats upward and collects on the underside of the crude oil released by the treatment. Plate 12.46 shows thread-like releases from the treated black goop.

Microscopic examination of debris perching under the floated crude oil (treatment water:volume ratio of 10:1) revealed typical structures found in bioconcretious growths, indicating that they were incorporated into the BAP consorm. Plate 12.47 reveals sheath-like structures similar to those found in sheathed IRB commonly

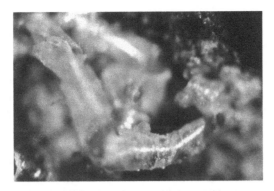

PLATE 12.47 Photomicrograph (×40) of voided sheath structures recovered from black "goop" dispersed by hot CB-4 treatment.

found under oxidative conditions on the edges of a biofouling biomass. Their presence in asphaltenes raises the question of where these strictly aerobic bacteria obtained oxygen in a very reductive environment. One possible answer is that the bacteria can manipulate electrical charges (cathodically impressed corrosion control) applied to flowing crude by creating bioanodic structures in the biomass. The release of oxygen from the water via bioelectrolysis would allow oxidative functioning to take place.

12.18   MONITORING METHODOLOGIES

Monitoring consormial activities should always be performed in the least invasive manner so the various cooperating components of a consorm are not impaired dramatically. Reductionism demands differentiation of a consorm to its most active component evidenced by isolation in a culture or a perceived genetic fingerprint also obtained from a culture. Consormial assessment of activities is possibly most easily achieved by examining some basic metabolic function. The presence of ATP is presently the best determinant of high energy metabolic function. The weakness (or perhaps strength) of the test is that it reveals total ATP at that moment when the samples was analyzed.

There is a natural contradiction between activity and presence for a consorm and for its participating component strains. Presence does not necessarily mean activity; activity is limited to cells that undertake a particular function. Any cultural procedure applied to a sample in which a consorm resides automatically becomes selective for cells that react positively in a cultural biochemical sense to the conditions presented. The simple act of storing a sample can cause changes in activity levels of various bacteria in a consorm. A genetic fingerprint of a consorm may be marred by nucleic acids and amino acids that may be present in the sample but not integral parts of a consorm. Culturing to refine a genetic fingerprint automatically produces distortion because intrinsic bias is always present. Precision shown when a procedure is repeated may simply mean the scientific errors are consistent. Precision does not correspond to accuracy in defining a consorm; it simply indicates repeatability.

Distortions are almost inevitable during identification and categorization. The failure to culture a consorm does not mean the consorm is not present; it simply means the cultural techniques are inadequate. Many bacterial pathogens remain to be discovered. They have not been cultured isolated simply because of two dogmatic assumptions: (1) pathogens all grow on agar media and are single strains; and (2) bacterial consorms cannot cause disease. Two major factors to be considered in the future are: (A) indigenous bacterial consorms can actually by competition prevent a clinical infestation of a host by pathogenic bacteria; and (B) some forms of cancer may actually be bacterial (strain or consorm) in origin. Bacteriologists need far better definitions of the roles and functions of bacterial consorms known at present.

Identifying a consorm involves observation of its form function and using a suitable ATP test to determine levels of metabolic activity. Another issue is the ability to culture fractions of the consorm in a manner that allows identification. BART analysis offers the advantage of generating two major gradients (selective nutrients diffusing upward; and oxidation–reduction gradient moving downward toward reductive); a broad spectrum of microniche environments appear in the testers. Selective BARTs define several major groups of bacteria that participate in consormial activities (Chapter 9). The sequenced approach is defined in Chapters 8 and 9; Chapter 8 addresses the biochemical aspects of identification. Chapter 9 details the various BART procedures.

Biochemical analysis of a consorm determines total ATP and extracts fatty acid methyl esters (FAME) using RASI-MIDI techniques to maximize the ability of a consorm to generate FAME as a chromatograph (fingerprint). The objective of RASI is to maximise aerobic cellular growth of a consorm within a selected BART in a manner that accelerates cell wall synthesis and production of FAME. RASI requires sequenced agitation under highly oxidative conditions followed by a static incubation period; both factors can be adjusted for specific consorm libraries.

Cultural identification of consorms using BART involves diluting a sample of the consorm into the specific tester and examining the reactions and time lapses (TLs) generated. This is best performed using the VBR72 system of TL photography that precisely reveals populations and reaction types. Reactions (see Tables 10.1 to 10.17 for details) are particularly important as are the times required for reactions, some as short as 1 hour.

Odd reactions can occur, particularly in the early stages of aerated treatment of sanitary wastewaters. Bacteria in a HAB BART vial can form slime threads (Plate 12.48) that gradually rotate and interconnect to the headspace at the water line on the BART ball. These threads remain oxidative (blue) when the surrounding

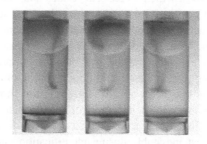

PLATE 12.48 Slime thread formations in primary influent sanitary wastewater samples revealed by HAB BART at 28°C.

culturing conditions are reduced (clear). The blue thread color results from the movement of oxygen from the headspace into the culturing fluids. Over time, the slime threads shorten and lose the ability to carry oxygen. As a result the extension distance decreases (arrow) and the activity weakens.

Large biomasses can accumulate within a tester charged with a consormial sample. Plate 12.49 shows (arrows) two SRB-BART testers inoculated with 0.5 g of 3, 19-21 PLG in 15 ml sterile distilled water. The right-hand testers were inoculated with a different patina that did not generate a large globular black biomass. The VBR72 system allows routine monitoring of reactions within individual testers (six testers per rack with four rows and three columns).

The BART rack (Plate 12.50) allows the precise positioning of the testers. The locks lock together to provide a stable platform for the VBR72 system. Each rack has a thick black line (arrow) that indicates cloudiness in a sample as a result of bacterial growth. Two thin lines are also printed on the backing screen to define more subtle changes resulting from growth. For use on ships and in moving vehicles, a supporting bar (white arrow) can be used to lock the testers in place.

PLATE 12.49   Unusual formation of black globular biomass under balls for two testers (left) using an OCR consorm recovered from patina growing on fiber-reinforced concrete pipe.

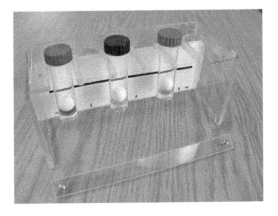

PLATE 12.50   BART rack showing three of the six positions loaded with testers and supporting bar used to reduce motion.

The visual BART reader (VBR) software system allows a single approved camera to take TL photographs every 15 minutes to allow interpretation of TLs and reactions that may be archived for future reference. Plate 12.51 shows the VBR48 system in that simultaneously monitors 48 testers in two columns of four rows of BART racks. Tests can be started at intervals. The software allows a specific start time to be entered for each tester and replaced when the tester shows a full positive reaction, the TL is entered, and the population predicted.

At present, BART can monitor a maximum of 72 testers (four rows × three columns × three replicates). Plate 12.52 shows HAB BART tests run from eleven sites in

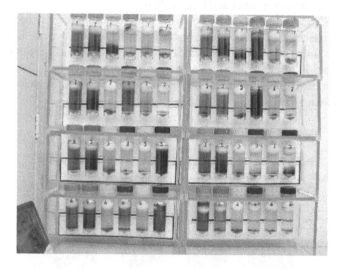

PLATE 12.51   VBR48 system in use for multiple BART analyses.

PLATE 12.52   VBR72 set to monitor eleven (replicated ×6) samples from different stages of wastewater treatment system using HAB BART incubated at 28°C.

an aerated wastewater treatment system incubated 28°C with digital images recorded every 5 minutes (six replicates of each stage of treatment from top left to bottom right). Reactions were always UP types and the first five sets of six tests were complete. The two lagoon samples (arrow indicating set of six) were becoming UP-positive while the bottom four sets remained negative. The VBR72 software recorded the reaction patter, noted TLs and, allowed interpretation by calculating populations. The VBR72 acts as a window to project the levels of specific bacteria in a sample.

Investigation and interpretation of bacterial consorms must consider the bacterial strains in a consorm. No single strain in a consorm is essential; the communities within a consorm can adapt to environmental conditions. They exchange genetic material as required and adapt it to maintain growth, form and function.

Suggestions for Further Reading

A generalistic approach to the classification of bacteria primarily as communities (consorms) as the first level to understanding and identifying them by "lumping" them together into logical groups makes common sense. One reviewer of a paper that I coauthored made an interesting statement: "Common sense is rarely used as hard evidence in scientific papers" but when working in the natural world (rather than the synthetic regime of a laboratory), common sense is a prerequisite for the development of a practical scientific hypothesis. Often in the practical application of science, it is a key ingredient.

Bacterial consorms as complex interactive communities are generated by local environmental conditions and are recognized by their forms, functions, and locations within which the associated biomass is generated. In such circumstances, away from a specialized laboratory, the common sense of the experienced observer becomes important. It is inevitable that lumpers must rely on experiences that are relevant to the perceived bacteriologically dominated growth. Reductionists, conversely, split the challenge down to the molecular and biochemical levels. The information generated, while precise, may not aid in the identification and management of a medical, environmental, or metallurgical condition. Basically, splitters define at high resolution while lumpers resolve at low resolution.

Over the past 30 years, a deluge of reductionist papers set the stage for the future development of bacteriology. Only a trickle of papers discussed the generalistic activities of bacterial consorms. Unfortunately the narrow spectrum of reductionist findings distorts the mirror of science. The readings listed below are in alphabetical order by order and provide sources for further investigation if desired:

Abu, W. et al. (2003). Assessment of the identification of diverse Helicobacter species, and application to faecal samples from zoo animals to determine Helicobacter prevalence. *J Med Microbiol* 52: 765–771.

This paper utilizes biochemical techniques for the refined identification of Helicobacterium species using PCR-DGGE. The atlas addresses this genus only as a part of 2, 06-13 ULC (ulcer). It is interesting that this genus was only recognized when blood agar plates were incubated for longer than the recognized standard 2 days.

Amann R.I., W. Ludwig, and K.H. Schleifer. (1995). Phylogenetic identification and in situ detection of individual microbial cells without cultivation. *Microbiol Rev* 59: 143–169.

This touches on a fundamental need to detect microbial cells without culture in situ. Such techniques, if sufficiently precise and economical, would go a long way toward defining the nature of bacteriological challenges.

Boone, D.R. and R.W. Castenholz. (2001). *Bergey's Manual of Systematic Bacteriology.* Volume 1: The Archaea and the Deeply Branching and Phototrophic Bacteria, 2nd ed. Baltimore: Williams & Wilkins.

Bergey's Manual, since the late 1920s, has presented the standard methods for differentiating bacteria, but it does not address the potential for complex consormial forms of bacterial growth. Volume 1 addresses the Archaea. See also Brenner et al. and de Vos et al. below for the other two volumes.

Brenner, D.J. et al. (2005). *Bergey's Manual of Systematic Bacteriology.* Volume 2: The Proteobacteria, 2nd ed. Baltimore: Williams & Wilkins.

Volume two concentrates on the gamma proteobacteria that include many of the alpha two bacterial consorms described in this atlas.

Costerton, J.W. (2007). *The Biofilm Primer,* Vols. 1–4. Heidelberg: Springer.

The past 30 years of research show that bacteriologically dominated biomass tends to be consormial in nature and can lead to various forms of biofouling. The premise is that these bacteria grow within biofilms that then interact in a natural or synthetic environment. Biofilms can mature to extended biomasses that totally fill voids and fractures, change water quality and flow, and lead to corrosion and collapses. Volume 1 of the series is a primer; volume 2 covers control of infections; volume 3 covers effects of biofilms; and volume 4 covers marine and industrial biofouling.

Cullimore, D.R. (2000). *Practical Atlas for Bacterial Identification,* 1st ed. Boca Raton: CRC/Lewis Publishers.

This atlas constituted the first attempt to place bacterial genera into a two-dimensional map, but did not address the potential for many bacterial genera to become integrated with other microorganisms into consormial structures.

Cullimore, D.R. (2008). *Practical Manual of Groundwater Microbiology,* 2nd ed. Boca Raton: CRC/Lewis Publishers.

Thirty years of field and laboratory investigations clearly show that biomasses generated within groundwater habitats are dominated by bacterial consorms, usually in association with a reductive–oxidative interface or some chemical nutrient stream entering infested voids or fractures. Emphasis is on common sense practices for controlling (managing) biofouling without dramatically affecting the purposes for which the groundwater was exploited.

de Vos, P. et al. (2010). *Bergey's Manual of Systematic Bacteriology.* Volume 3: The Firmicutes, 2nd ed. Baltimore: Williams & Wilkins.

This volume does not address the community structures (consorms) involving different genera integrated into a common dynamic biomass. These consorms intelligently respond to shifts in the local environment.

Dyer, B.D. (2003). *A Field Guide to Bacteria.* Carson City, NV: Comstock Books. This field guide is a general introductory text.

Falkow, S. et al., Eds. (2007). *The Prokaryotes.* Volume 1: Symbiotic Associations: Biotechnology and Applied Microbiology; Volume 2: Ecophysiology and Biochemistry; Volume 3: Archaea and Bacteria: Firmicutes and Actinomycetes; Volume 4: Bacteria: Firmicutes and Cyanobacteria; Volume 5: Proteobacteria: Alpha and Beta Subclasses; Volume 6: Proteobacteria: Gamma Subclass; Volume 7: Proteobacteria: Delta and Epsilon Subclasses: Deeply Routing Bacteria; 3rd ed. New York: Springer-Verlag.

The Prokaryotes continues to challenge *Bergey's Manual.* The third edition published in 2007 contains over 7,000 pages. Both publications share a common basis, but *The Prokaryotes* pursues a more defined reductionist approach. It does not adequately address the potential for bacteria to form consorms.

Ellis, D. (1919). *Iron Bacteria*. London: Methuen & Co.

This 1919 book was lost on the dusty shelves of libraries and almost forgotten until it was republished in 2006. David Ellis and colleagues wrote a very descriptive, practical book about iron bacteria. His classification scheme is still in place today and his book describes the role of iron bacteria in the formation of pig iron. Today 3, 19-16 RST consorms generate pig iron of various grades on steel shipwrecks in the deep-ocean environment. David's book is a required read because it includes field experience, common sense, and laboratory verification.

Michael, F. (2003). Accurately identifying bacteria, *Anal Chem*. 75: 143.

This article explores the need to accurately identify bacteria. It also tends to divide and conquer—separate bacteria into forms that can then be identified.

Figuerola, E. and Erijman, L. (2007). Bacterial taxa abundance pattern in an industrial wastewater treatment system determined by the full rRNA cycle approach. *Environ Microbiol*, 9: 1789–1789.

This article describes the abundance of different bacteria within industrial wastewater by looking at the full rRNA cyclic approach and potential identification of bacteria in managed ecosystems.

Gong, J. et al. (2002). Diversity and phylogenetic analysis of bacteria in the mucosa of chicken ceca and comparison with bacteria in the cecal lumen. *FEMS Microbiol Lett*, 208: 1–7.

This descriptive analysis of bacteria within chicken ceca and cecal lumen emphasized the diversity determined by phylogenetic analysis that involves the reductionist approach toward the determination of bacterial communities in chickens.

Kunitsky, K. et al. (2006). Identification of microorganisms using fatty acid methyl ester (FAME) analysis and the MIDI Sherlock microbial identification system, in Miller, M.J., Ed., *Encyclopedia of Rapid Microbiological Methods*, Vol. 1, Bethesda, MD: PDA Publications, pp. 1–17.

Kunitsky describes major biochemical pathways for identifying bacteria are nucleic acid analysis and determination of fatty acid compositions in cell walls. Chapter 8 of this atlas describes the RASI-MIDI methodology for identifying bacterial consorms using method S43020.

Little, B.J. and J.S. Lee. (2007). *Microbiologically Influenced Corrosion*. New York: John Wiley & Sons. This book is a basic text covering economically significant corrosive processes that impact pipelines, wells, industrial plants, and distribution and storage systems. Emphasis is on processes that cause various forms of corrosion in steels, metal alloys, and concretes.

McGill, B.J. et al. (2007). Species abundance distributions: moving beyond single prediction theories to integration within an ecological framework, *Ecol Lett*, 10: 995–1015.

This article addresses some of the issues involved in attempting to integrate predictive challenges into a modelling exercise. Determining species abundance is perhaps the first step toward recognizing the dynamic nature of consorms.

Muyzer, G. et al. Phylogenetic relationships of Thiomicrospira species and their identification in deep-sea hydrothermal vent samples by denaturing gradient gel electrophoresis of 16S rDNA fragments. *Arch Microbiol*, 164: 165–72.

Deep-sea hydrothermal vents are complex sites where great numbers of environmental reactions occur along thermal, ORP, and water chemistry gradients. This article presents a method for identifying *Thermomicrospira* species using 16SrDAN fragments.

Narang, R. and Dunbar, J. (2004). Modeling bacterial species abundance from small community surveys. *Microb Ecol*, 47: 396–406.

Understanding bacterial species abundance in small community surveys presents the challenge of modeling such events. Modeling is the main subject.

Osborn, A.M. et al. (2000). An evaluation of terminal restriction fragment length polymorphism (T-RFLP) analysis for the study of microbial community structure and dynamics. *Environ Microbiol*, 2: 39–50.

This article evaluates attempts to use T-RFLP analysis to study microbial community structures. The technique may lead to improved identification techniques for bacterial communities and consorms.

Passman, F. (2003). Fuel and fuel system microbiology: fundamentals, diagnosis, and contamination control, *ASTM Manual 47; IASH Newsletter*, Issue 29.

Corrosion is a significant issue in hydrocarbon-based fuels, particularly during storage and conveyance of petroleum-based product. A combination of steel surfaces, water, and fuel can trigger bacteriologically influenced contamination of the fuel, rendering it useless for energy production and generating focal sites for corrosion. ASTM Manual 47 explains methodologies and control methods to minimize these events.

Postgate, J. (2000). *Microbes and Man*, 4th ed. Cambridge: Cambridge University Press.

John Postgate wrote four editions of *Microbes and Man* and they were all easy reading. They addressed the challenges that microbes presented to human society. Ideally, all four books should be read in sequence to gain an understanding of the changes in microbiology over the past 50 years.

Stead D.E. et al. (1998). Modern methods for identifying bacteria, *Plant Cell Tiss Organ Cult*, 52: 17–25.

Cell cultures from plant, tissue and organs can also contain bacteria. Stead reviews the methods for identifying bacteria within such cultures.

Vesela, A., Tzeneva, H., Hellig, G.H.J., Van Vliet, W.A., and D.L. Atoon. (2007). 16S rRNA targeted DGGE fingerprinting of microbial communities, Env Genomics, 410, 335–350.

Attempts have been made to fingerprint microbial communities using 16S rRNA targeted DGGE. This is a reductionist approach using the narrower basis of selected genetic markers.

Zhang, H. and T.A. Jackson. (2008). Autochthonous bacterial flora indicated by PCR-DGGE of 16S rRNA gene fragments from the alimentary tract of Costelytra zealandica (Coleoptera: Scarabaeidae), *J Appl Microbiol*, 105: 1277–1285.

This paper utilizes PCR-DGGE analysis of the 16SrRNA gene fragments to define bacteria found within insect guts.

Appendix
Alpha Two Traditional
Atlas Concept

The first edition of the *Practical Atlas for Bacterial Identification* by D. Roy Cullimore was published in 2000 and went on to become a best seller. At that time bacteriologists were preoccupied with what are now in the second edition referred to as alpha two organic bioconcreting bacteria. Historically, most of the progress focused on the alpha two bacteria differentiated at the section and genus levels in the first edition. In the past 10 years, bacteriological meanderings have taken us to many different arenas where bacteria play important roles. In the second edition, the basic concept is expanded from the reductionist mentality of following the Linnaean practices of zoology and botany that differentiate genus and species to a method that examines bacterial communities (consorms) as the primary drivers of bacterial activities.

The second edition moves beyond the Linnaean concepts that essentially limit classification to organic synthesizing bacteria and proposes that bacterial communities are differentiated into six alpha groupings based upon their environment and the nature of generated daughter products. In the first edition of the atlas, these community structures were not considered so significant. This appendix reiterates the old concepts used essentially for the differentiation of alpha two bacteria.

The first edition graphically illustrated differentiations of the major groups of bacteria within the traditional concepts. In the first edition these concepts were introduced, but no attempt was made to interrelate these sections to the alpha consorm concept developed in the second edition. The format of the first edition followed the two-dimensional concepts commonly used by geographers for mapping. Instead of countries, the units are sections involving families and then genera. In developing the first edition of the atlas, a number of basic considerations to set the concepts were used (but not explained). The earlier concepts employed differentiation to genus level only for alpha two bacteria and the system was not set in the grid format based on oxidative–reductive potential (expressed in factorial millivolts) and water viscosity (expressed in log centipoise). Section and genus classifications were based on evidence of positioning related to neighboring groups of microorganisms.

Figure A.1 shows the interrelationships of alpha two bacteria and related groups of algae, fungi, and protozoa. The traditional concept positioned genera as able to function oxidatively (aerobically) and/or reductively (anaerobically), as shown in Figure A.2. One major factor that differentiates alpha two bacteria is the formation of cell walls from a relatively simple, unstructured (Gram-negative) form to a lateralized and complexed (Gram-positive) form, as illustrated in Figure A.3.

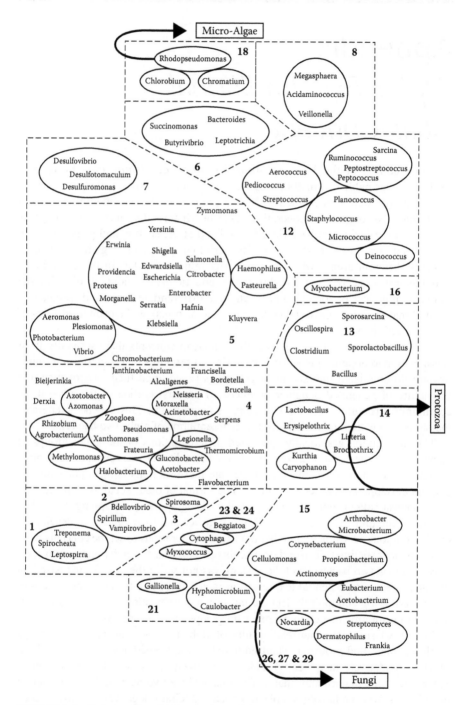

FIGURE A.1   Relationship of alpha two bacterial sections to other microorganisms grouped as microalgae, fungi, and protozoa.

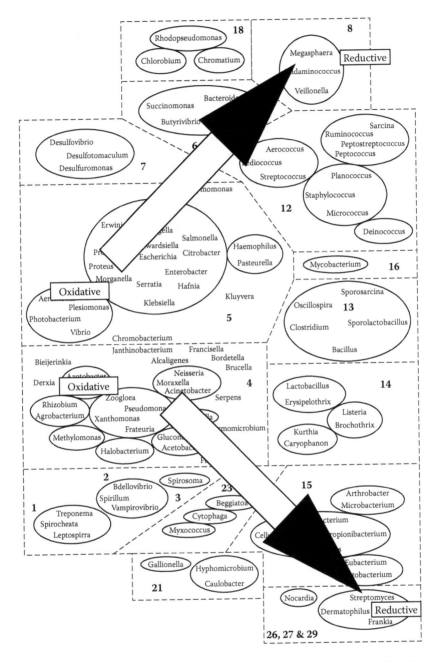

FIGURE A.2 Relationship of oxidative (open parts of arrows) and reductive (closed parts of arrows) conditions indicating bacterial sections that are predominantly aerobic (oxidative) and those that are anaerobic (reductive). See also Figures A.9 through A.11 for more details.

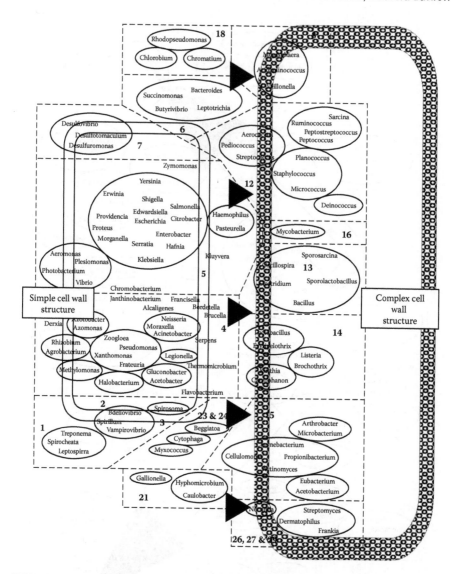

FIGURE A.3 Differentiation of bacterial sections based on the complexities of cell wall structures from relatively simple forms (Gram-negative, open-walled rectangle) to more complex structures (Gram-positive, shaded walled rectangle). See also Figures A.5 and A.6 for differentiation using Gram stain methodology.

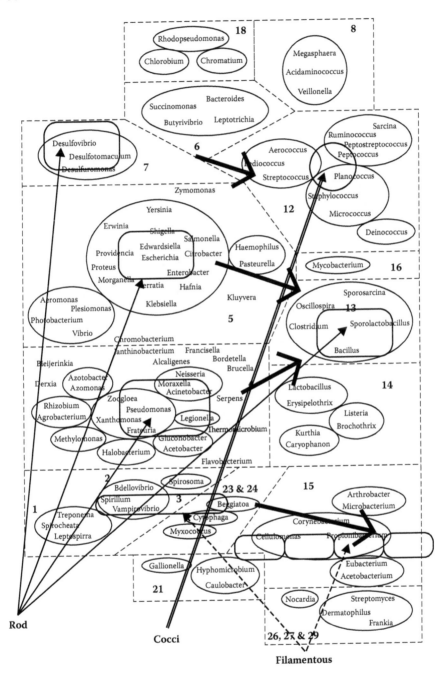

FIGURE A.4 Differentiation of bacterial sections based on cell shape. Basic forms are coccoid (spherical), rod-shaped, and filamentous rod forms.

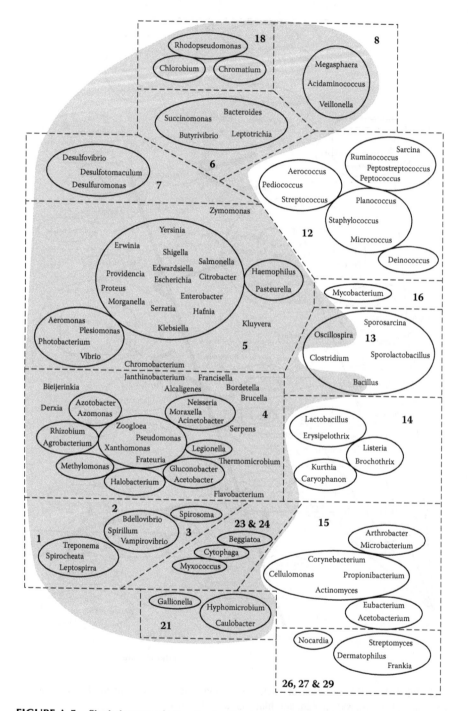

FIGURE A.5 Shaded zones of atlas display Gram-negative bacterial genera that have relatively simple structures within cell walls.

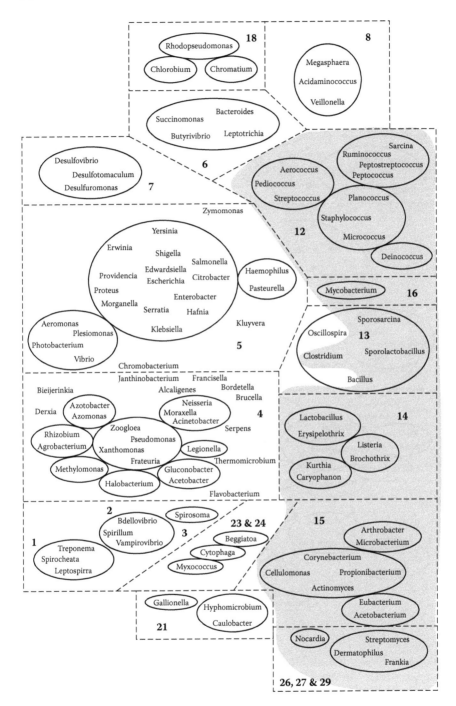

FIGURE A.6 Shaded zones of atlas display Gram-positive bacterial genera that have more complex cell walls.

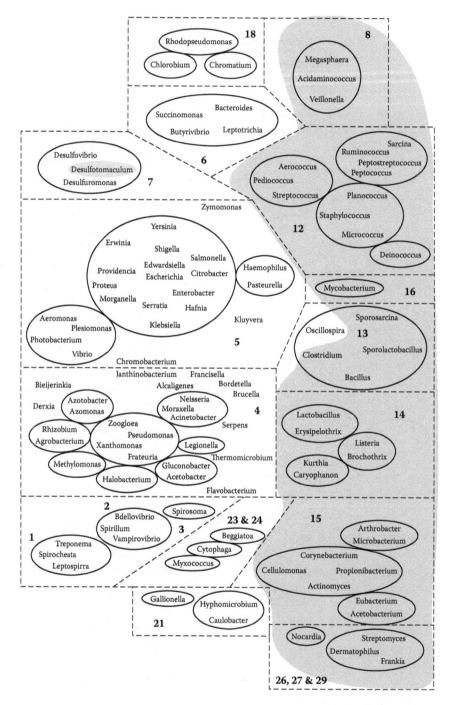

FIGURE A.7 Shaded zones of traditional alpha two atlas display bacterial genera commonly associated with Firmicutes, Division A (based on *The Prokaryotes*, second edition).

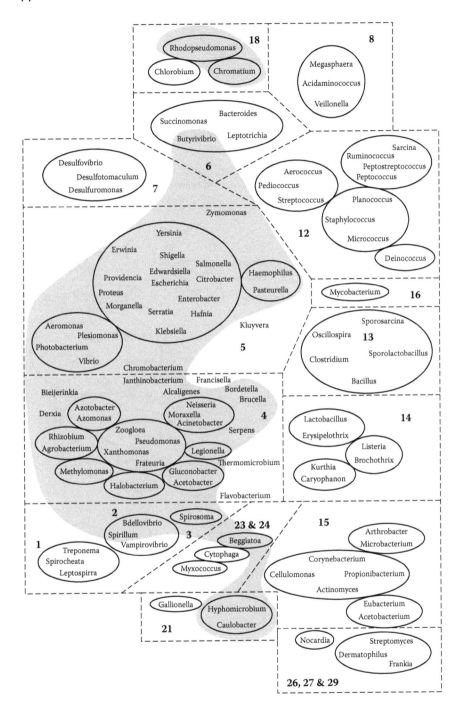

FIGURE A.8 Shaded zones of traditional alpha two atlas display bacterial genera commonly associated with Proteobacteria, Division C (based on *The Prokaryotes,* second edition).

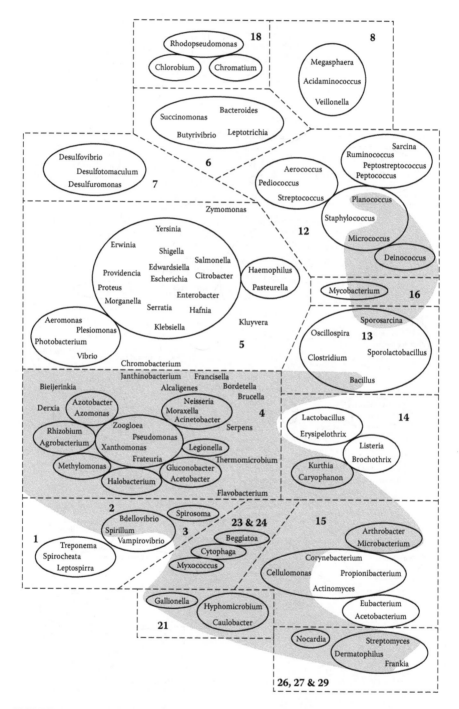

FIGURE A.9 Shaded zones of traditional alpha two atlas display bacterial genera that are strictly aerobic.

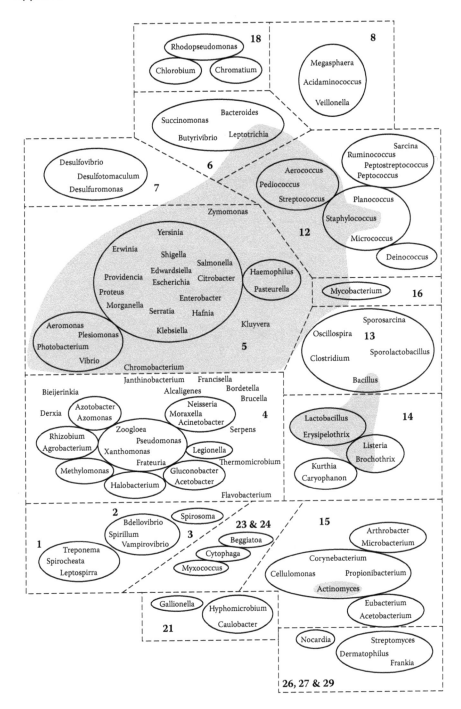

FIGURE A.10 Alpha two bacterial consorms set in traditional atlas format. Shaded zones indicate bacterial genera that are facultatively anaerobic.

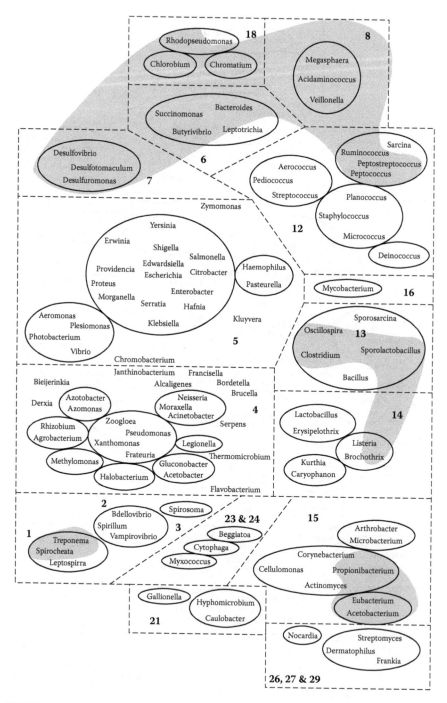

FIGURE A.11 Bacterial genera that are strictly anaerobic are displayed as shaded zones in traditional atlas format.

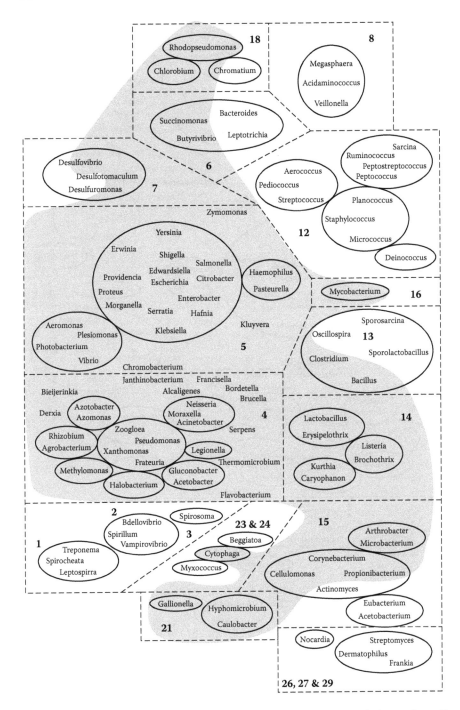

FIGURE A.12 Bacterial genera that are rod-shaped are displayed as shaded zones in traditional alpha two atlas format.

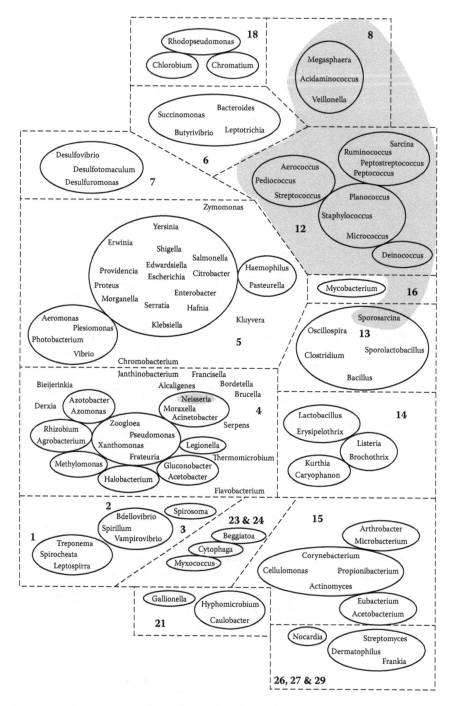

FIGURE A.13 Bacterial genera that are coccoid-shaped are displayed as shaded zones in traditional alpha two atlas format.

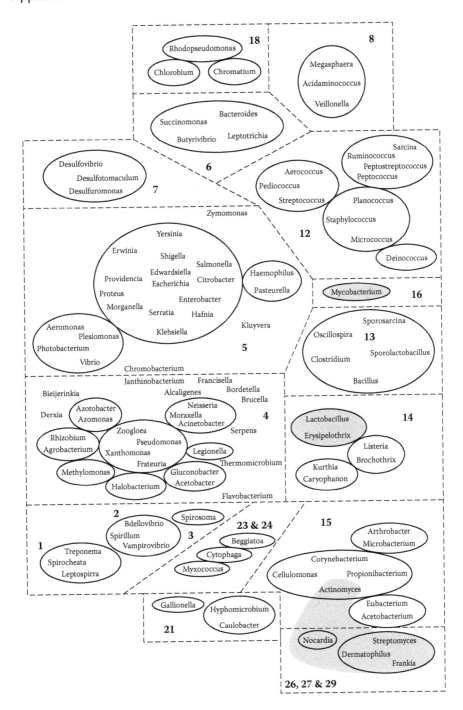

FIGURE A.14 Bacterial genera that are extensively filamentous (mycelia) are displayed as shaded zones in traditional alpha two atlas format.

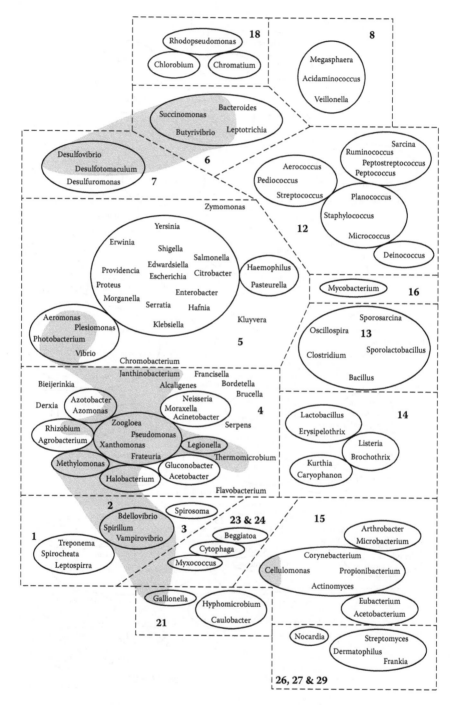

FIGURE A.15 Polar flagellation (shaded zones) for bacterial genera is shown in traditional alpha two atlas format.

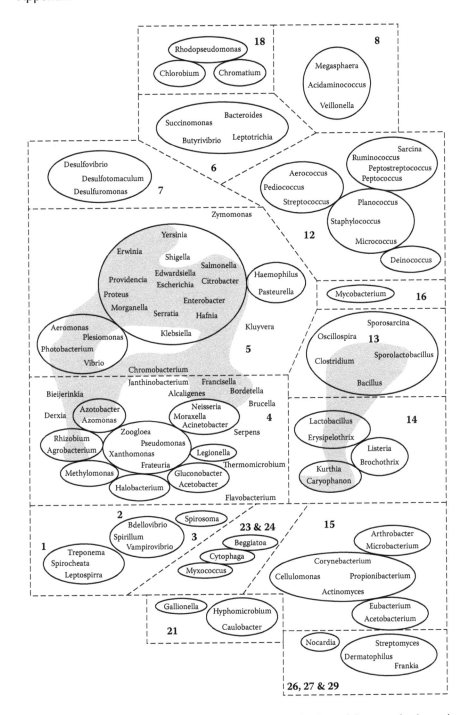

FIGURE A.16 Peritrichous flagellation (shaded zones) for bacterial genera is shown in traditional alpha two atlas format.

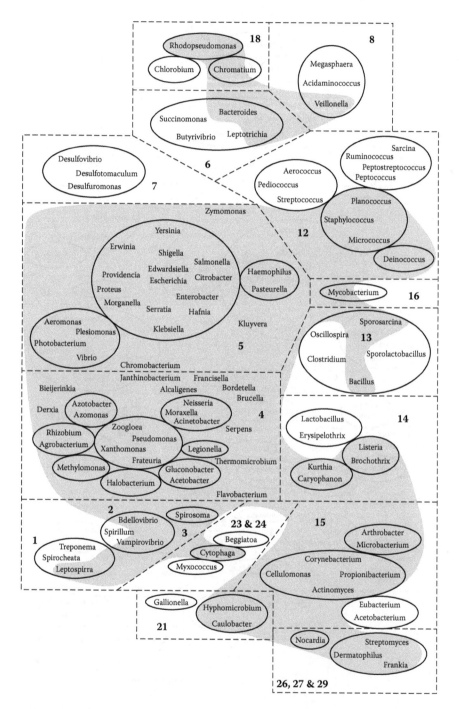

FIGURE A.17 Bacterial genera that are catalase-positive and aerobic (shaded zones) are shown in traditional alpha two atlas format.

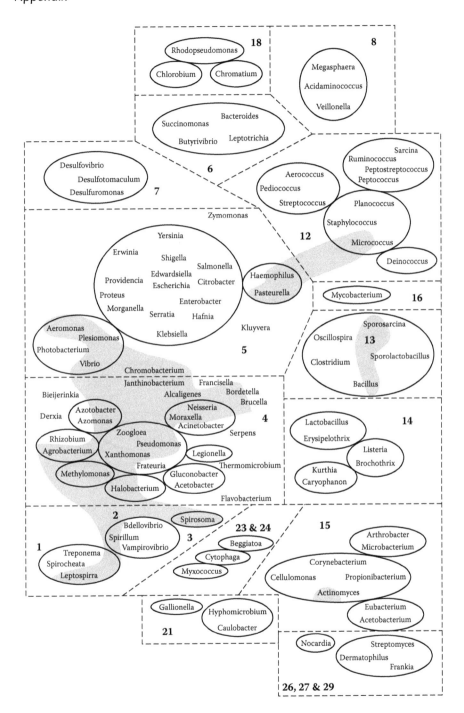

FIGURE A.18 Bacterial genera that are oxidase-positive (shaded zones) are shown in traditional alpha two atlas format.

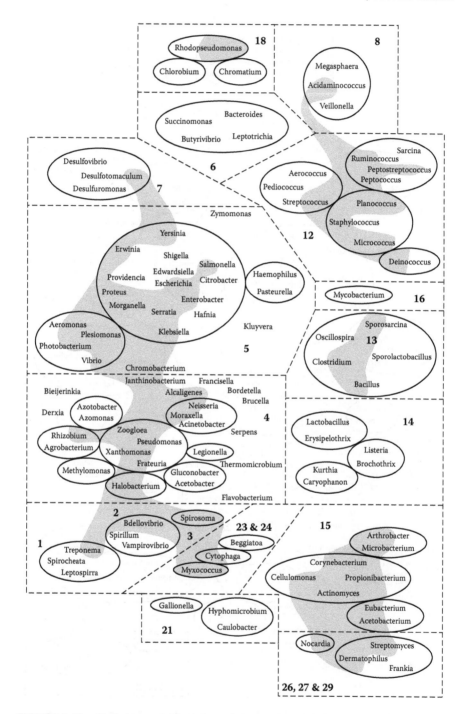

FIGURE A.19 Shaded zones indicate bacterial genera that can degrade proteins (are proteolytic) in traditional alpha two atlas format.

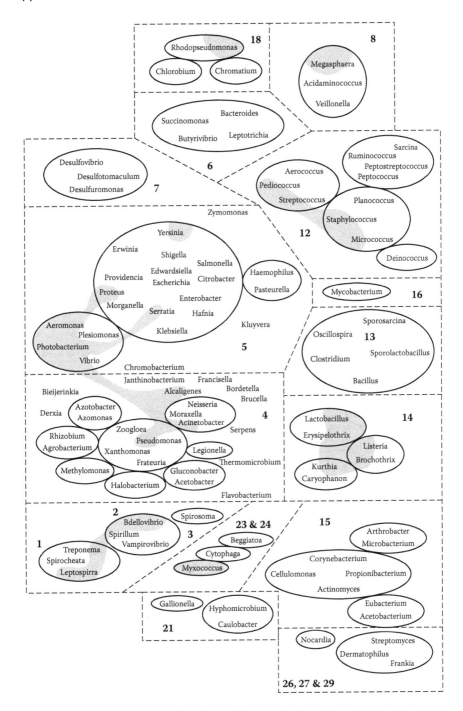

FIGURE A.20 Shaded zones indicate bacterial genera that break down fats (are lipolytic) in traditional alpha two atlas format.

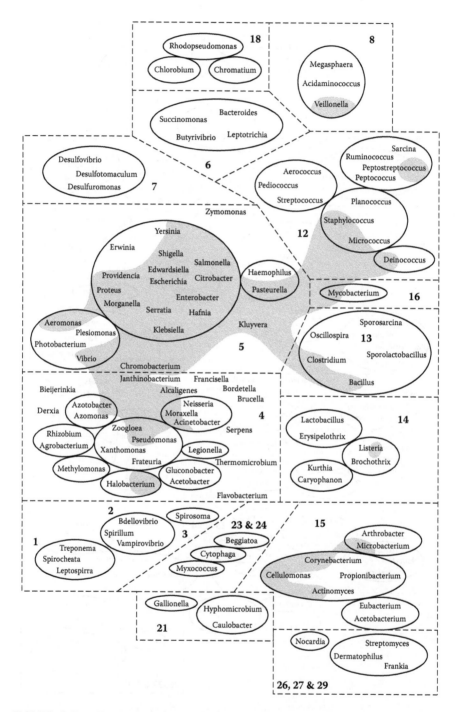

FIGURE A.21 Many aerobic bacteria also respire by denitrifying (reducing nitrate to nitrite). Shaded zones in traditional alpha two atlas format display bacterial genera that respire or otherwise reduce nitrates.

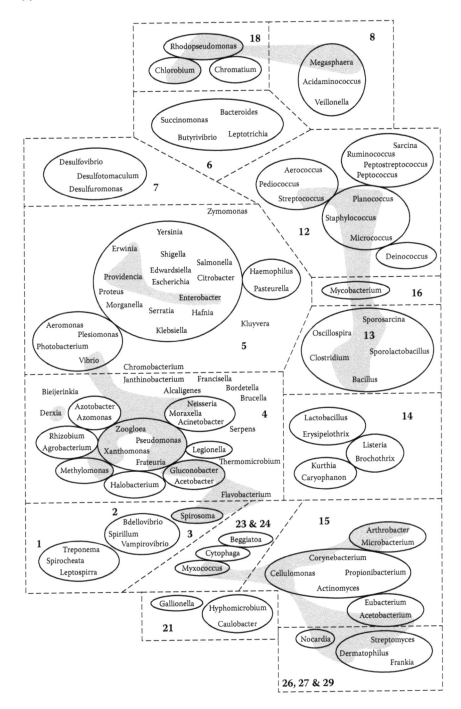

FIGURE A.22 Some bacterial genera frequently exhibit yellow pigmentation, particularly later during growth; they appear as shaded zones in traditional alpha two atlas format.

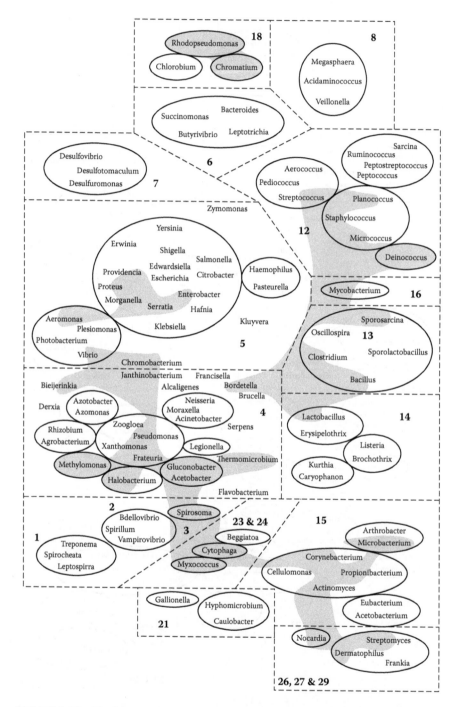

FIGURE A.23 Traditional alpha two atlas format displays (shaded zones) bacterial genera that exhibit red, orange, or pink pigmentation, particularly during maturation and growth.

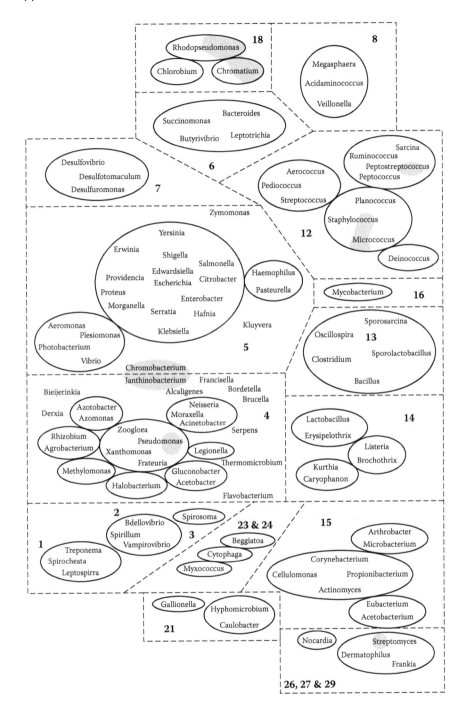

FIGURE A.24 A few bacterial genera exhibit purple or violet pigmentation, particularly later during maturation and growth; they appear as shaded zones in traditional alpha two atlas format.

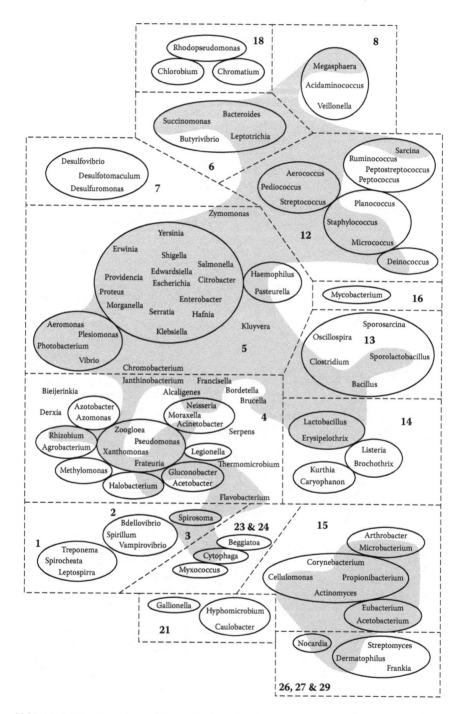

FIGURE A.25 Many bacterial genera degrade glucose and generate acidic end products as universal organic carbon sources for degradation and energy. Traditional alpha two atlas format shows them as shaded zones.

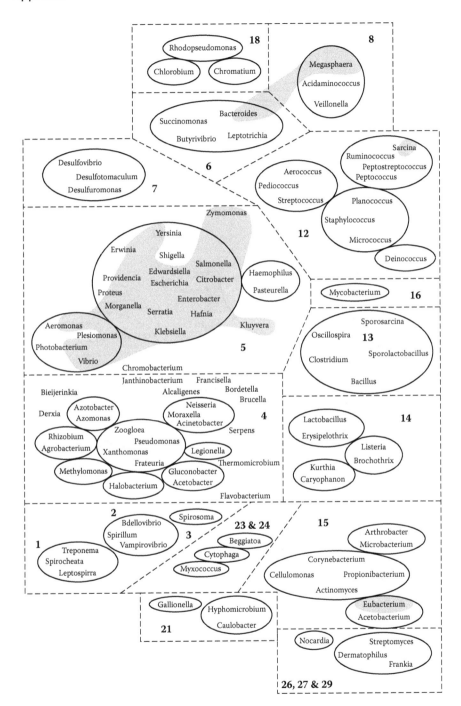

FIGURE A.26 Traditional alpha two atlas format displays (shaded zones) bacterial genera that degrade glucose and generate gaseous end products (mainly carbon dioxide and hydrogen).

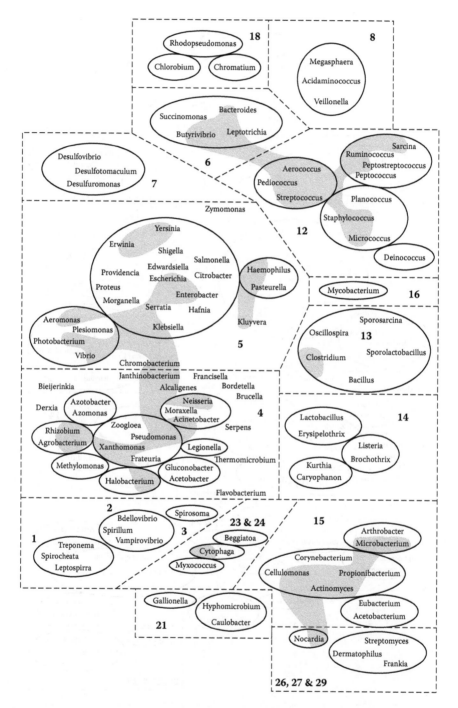

FIGURE A.27   Bacterial genera that ferment lactose (common sugar in mammalian milk) to acidic daughter products are displayed in traditional alpha two atlas format.

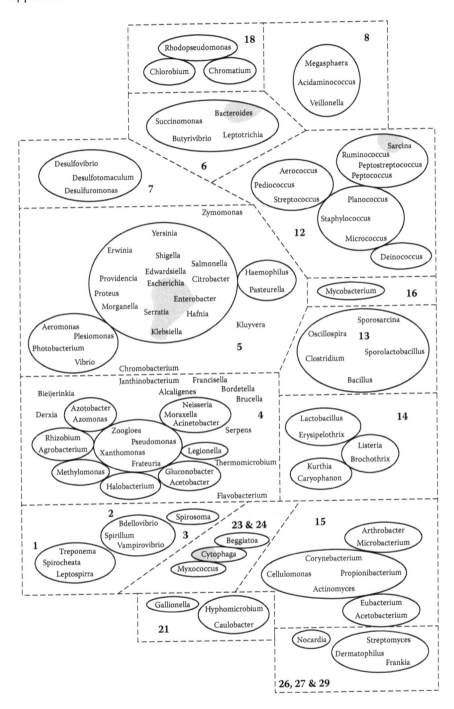

FIGURE A.28 Lactose can also be degraded via production of gaseous end products (mainly carbon dioxide and hydrogen) and acids. These are shown as shaded zones in the traditional alpha two atlas format.

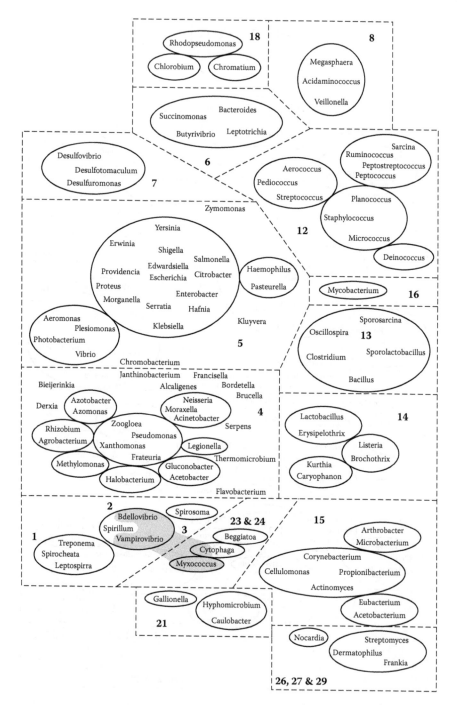

FIGURE A.29 Shaded zones in traditional alpha two atlas format indicate bacterial genera that act as parasites on other microorganisms.

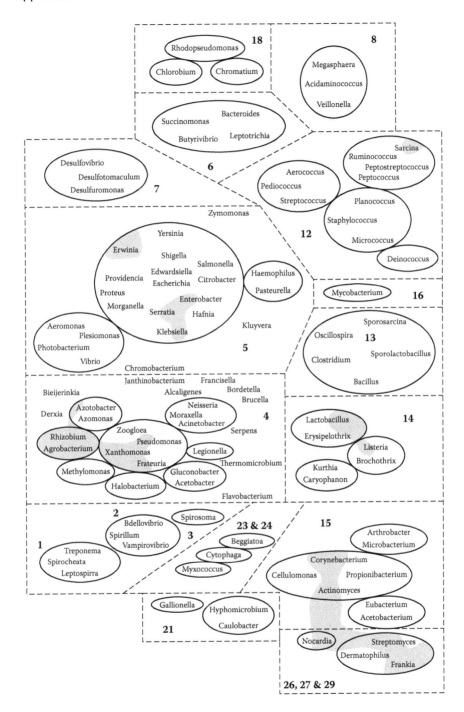

FIGURE A.30 Shaded zones in traditional alpha two atlas format indicate bacterial genera that act as parasites on plants.

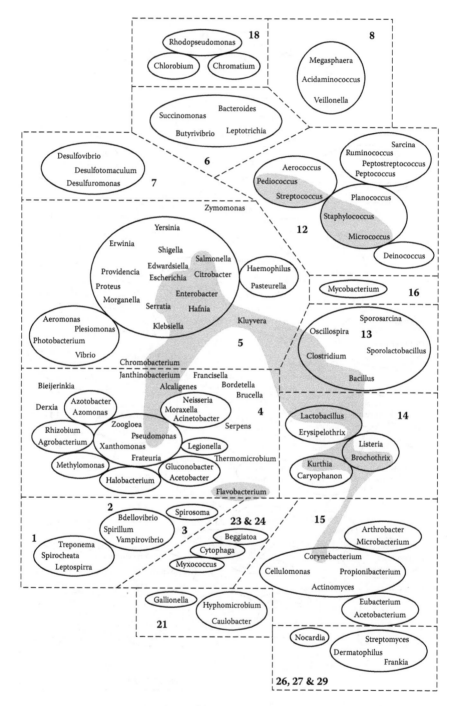

FIGURE A.31 Shaded zones in traditional alpha two atlas format indicate bacterial genera that cause spoilage in milk and dairy products.

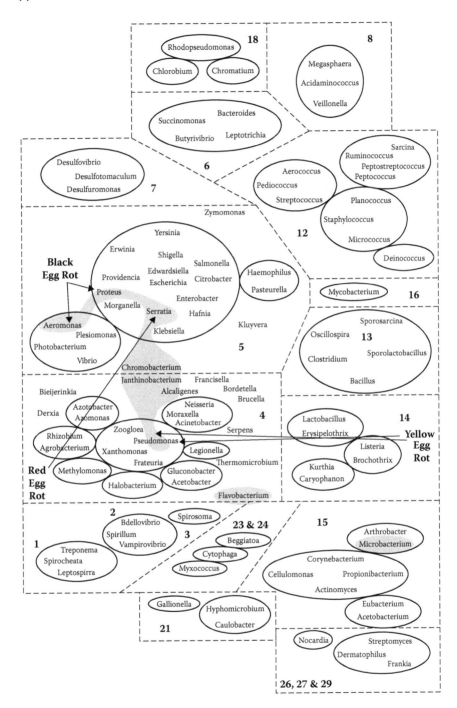

FIGURE A.32 Shaded zones in traditional alpha two atlas format indicate bacterial genera that cause spoilage in egg and poultry products.

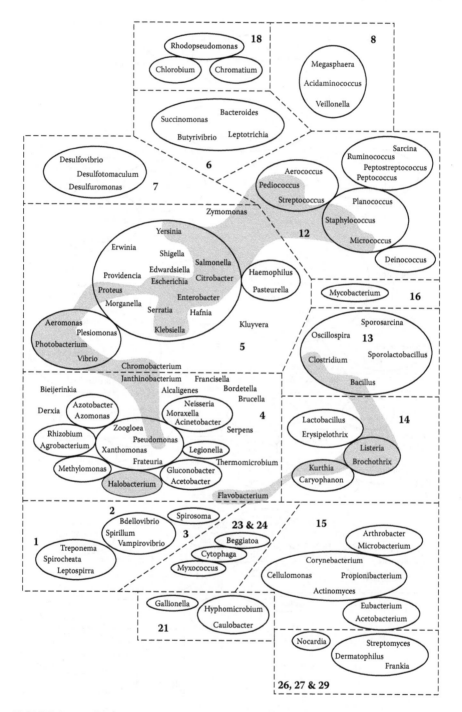

FIGURE A.33 Shaded zones in traditional alpha two atlas format indicate bacterial genera that cause spoilage in meat products.

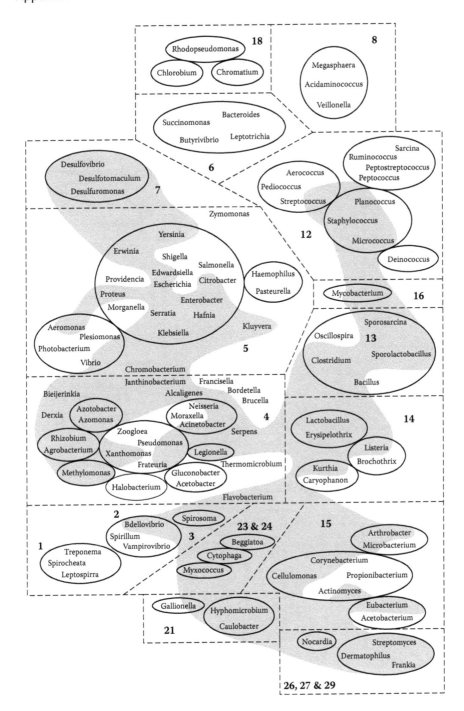

FIGURE A.34 Shaded zones in traditional alpha two atlas format indicate bacterial genera that include species known to commonly inhabit soils.

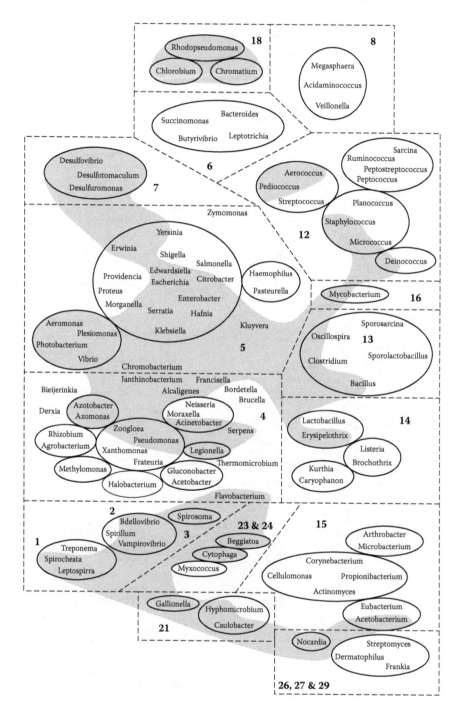

FIGURE A.35 Shaded zones in traditional alpha two atlas format indicate bacterial genera that include species known to commonly inhabit water.

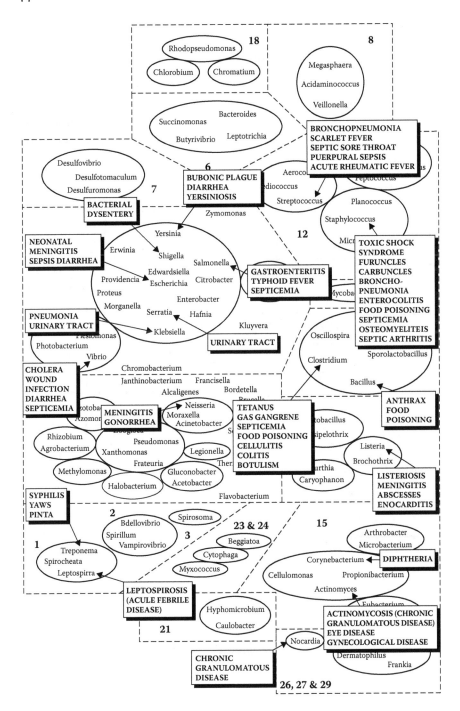

FIGURE A.36 Shaded zones represent bacterial genera that include species known to cause infectious diseases in humans (see partial listings in Figures A.37 and A.38).

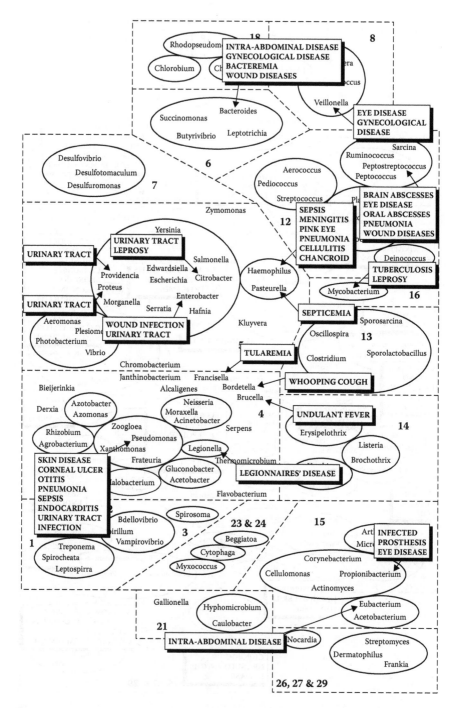

FIGURE A.37 Shaded zones represent bacterial genera that include species known to cause infectious diseases in humans (see partial listings in Figures A.36 and A.38).

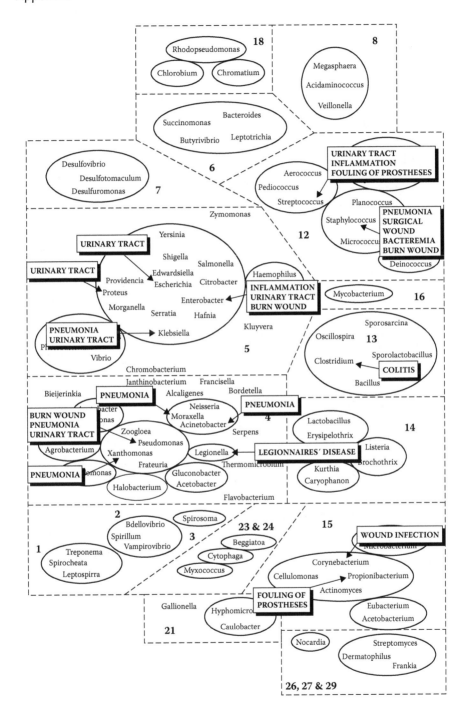

FIGURE A.38 Shaded zones represent bacterial genera that include nuisance and nosocomial species known to cause infectious diseases in humans.

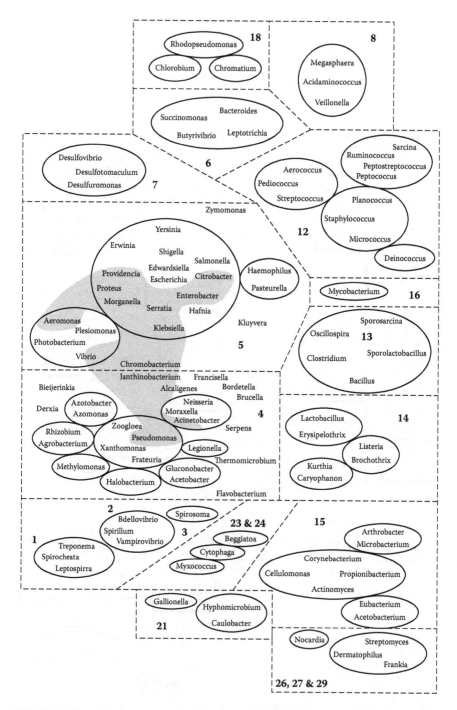

FIGURE A.39 Common bacterial consortial complexes in iron-related bacteria (IRB) BART that cause an RPS involving the reaction codes FO–GC–BL in sequence are likely to contain many genera highlighted in the atlas sheet above; section 4 pseudomonad bacteria dominate the consorm.

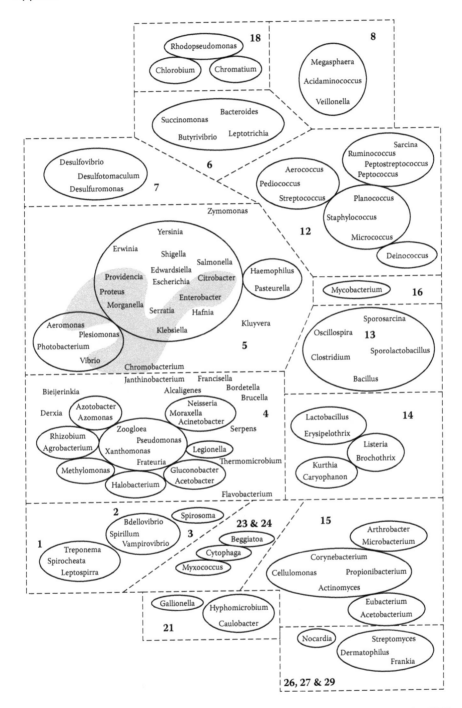

FIGURE A.40 Common bacterial consortial complexes in iron-related bacteria (IRB) BART that cause an RPS involving the reaction codes FO–CL–RC in sequence are likely to contain many genera highlighted in the atlas sheet above; section 5 enteric bacteria dominate the consorm.

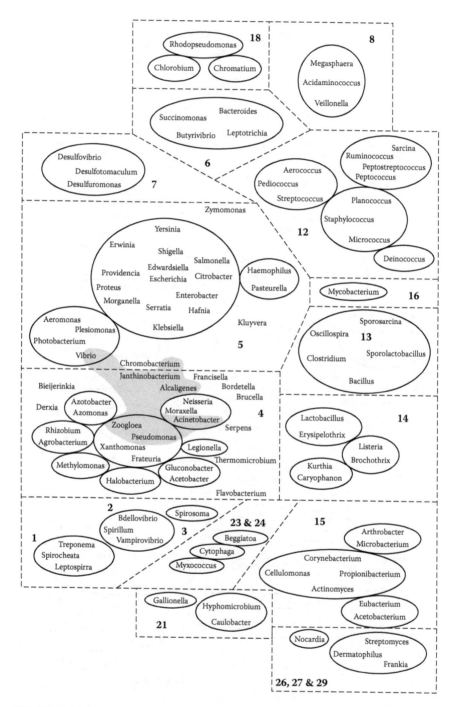

FIGURE A.41 Common bacterial consormial complexes in iron-related bacteria (IRB) BART that cause an RPS involving the reaction code GC possibly preceded by a CL.

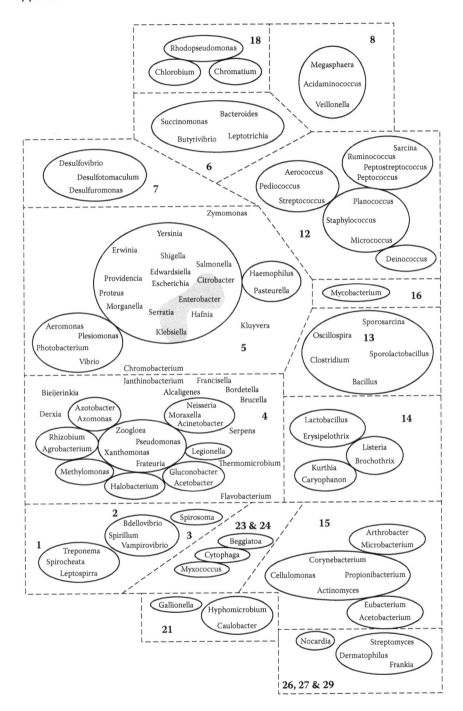

FIGURE A.42 Common bacterial consortial complexes in iron-related bacteria (IRB) BART that cause an RPS involving the reaction codes CL–BG in sequence are likely to contain many genera highlighted in the atlas sheet above; section 5 enteric bacteria including Enterobacter dominate the consorm.

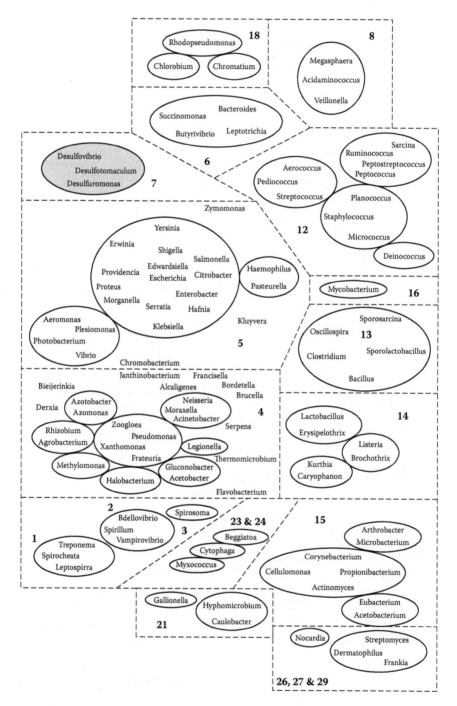

FIGURE A.43 Common bacterial consortial complexes dominated by sulfate-reducing bacteria (SRB). Note that Desulforuromonas are present only when an elemental sulfur substrate is available.

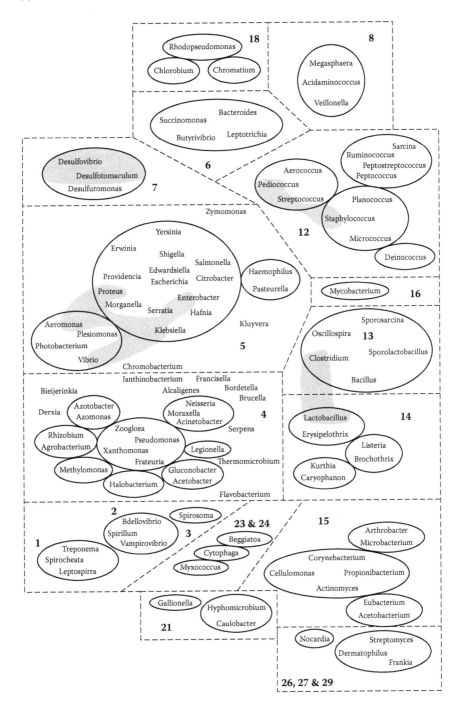

FIGURE A.44   Common bacterial consortial complexes in sulfate-reducing bacteria (SRB) are dominated by section 7 SRBs. In the BB reaction in which blackening appears in the base of the test vial, the likely genera present are shown in the atlas above.

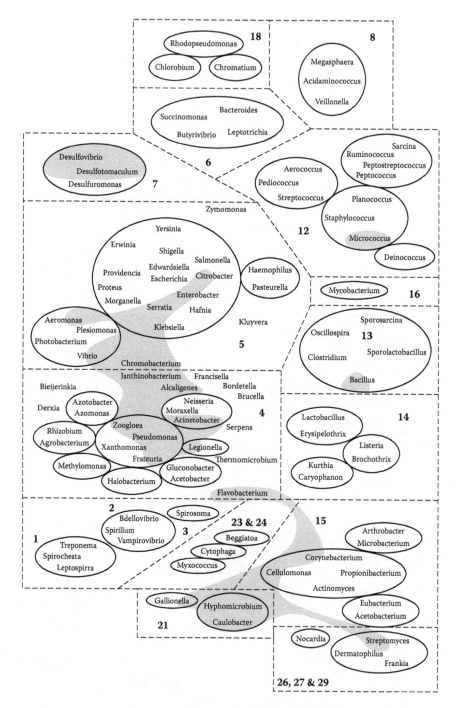

FIGURE A.45 Sulfate-reducing bacteria (SRB) BART generates a BT reaction (shaded zone) in which blackening often forms often in a granular manner around the ball in the test vial; the likely genera present are shown in the atlas above. SRBs are involved in primarily aerobic consortia and survive in the deeper layers of biofilms.

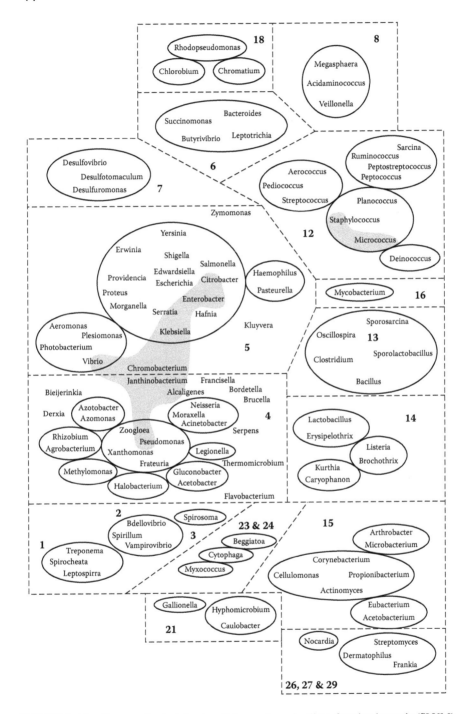

FIGURE A.46 Common bacterial consortial complexes in slime-forming bacteria (SLYM) BART are dominated by bacteria that generate copious slime (EPS) formations. The DS–CL reaction pattern triggered by enteric bacteria (e.g., Enterobacter and Klebsiella) can generate dense slimes followed by cloudiness.

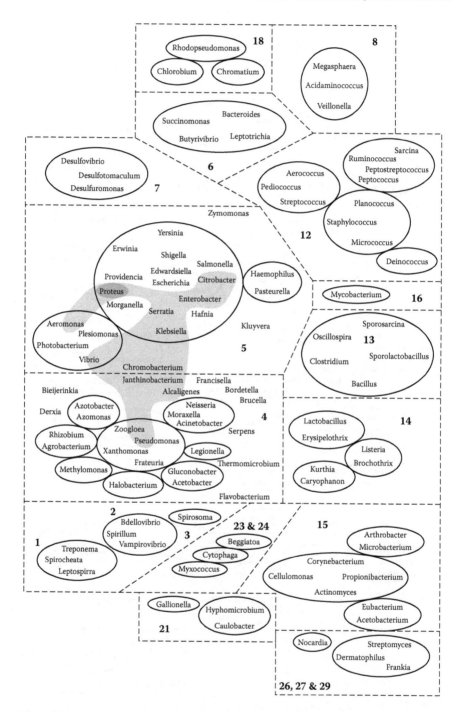

FIGURE A.47 Common bacterial consormial complexes in slime-forming bacteria (SLYM) BART are dominated by bacteria that generate slime (EPS) formations generating CP–CL reaction patterns.

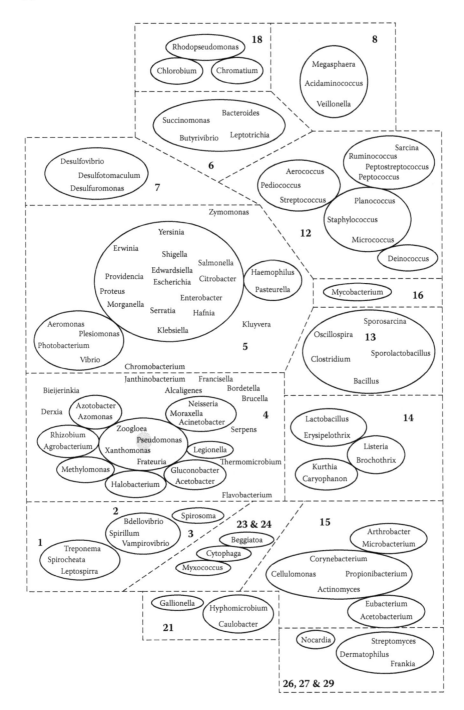

FIGURE A.48 Common bacterial consormial complexes in slime-forming bacteria (SLYM) BART are dominated by bacteria that generate slime (EPS) formations. In the CL–PB and CL–GY reaction patterns, Pseudomonas is the dominant genus in the consorm.

Another factor of primary interest is the variation of cell shape from coccoid (spherical), to short rod, long rod, and curved forms shown in Figure A.4. The remaining figures follow the same order presented in the first edition. These include the Gram reaction defined by the nature of cell wall structures that are Gram-negative (Figure A.5). Figures A.6 and A.7 delineate divisions used classically in the differentiation of Gram-positive bacteria (*Bergey's Manual*), then using the concepts employed for Firmicutes in Division A of the prokaryotes. Figure A.8 depicts the concepts employed for Firmicutes in Division B of the prokaryotes.

Aerobicity (Figures A.9 through A.11) is illustrated for strictly aerobic, facultatively anaerobic, and anaerobic bacteria. Cell shapes are illustrated as rods, cocci, and filaments in Figures A.12 through A.14. Figures A.15 and A.16 portray two main forms of motility involving polar movements (occurring at the ends of cells) and peritrichous movements (directed all around cells) as whip-like flagellae.

The traditional alpha two atlas addresses genera possessing specific enzyme systems including catalase (breaking down peroxides, Figure A.17), oxidase (Figure A.18), proteases (breaking down proteins, Figure A.19), and lipases (breaking down fats, Figure A.20). Genera able to denitrify (reduce nitrate to nitrite) are displayed in Figure A.21.

Figures A.22 through A.24 show the range of bacterial genera that can produce pigments of specific colors—yellows, reds, pinks, and violets. Genera that degrade glucose and lactose to acidic or gaseous products appear in Figures A.25 through A.28. The ranges of bacterial genera that parasitize other microorganisms or plants are shown in Figures A.29 and A.30, respectively.

Bacterial spoilage of foodstuffs is a major economic and hygienic concern. Figures A.31 through A.33 show genera that cause spoilage in dairy, poultry, and meat products, respectively. Figures A.34 and A.35 depict common alpha two bacterial genera found in soils.

Figures A.39 through A.41 deal specifically with bacterial genera that cause diseases in humans and indicate genera responsible for common diseases in tabular form by genus. The remaining figures define the generic compositions of various bacterial consortia detectable by the BART system (Figures A.42 through A.48).

Index